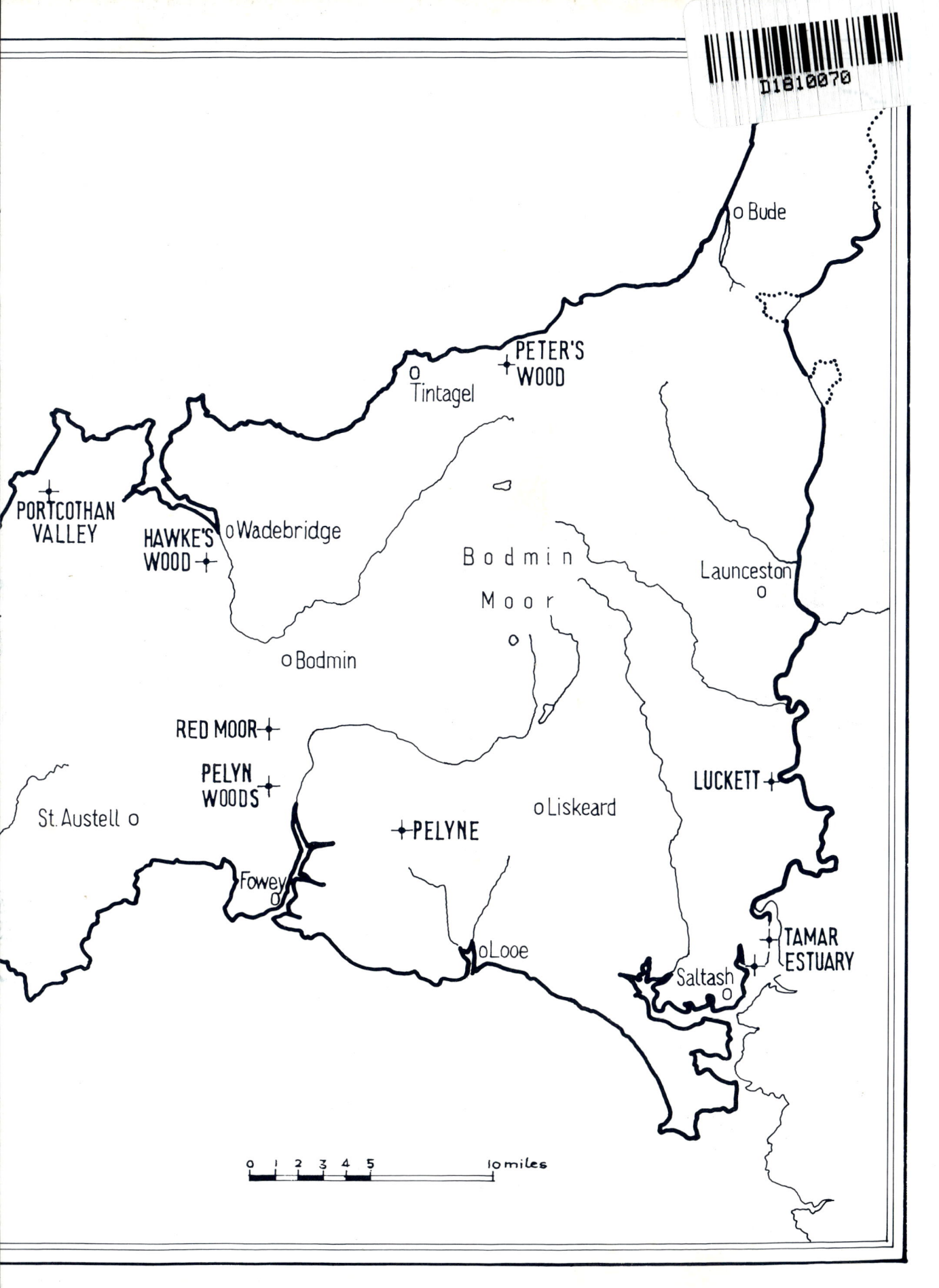

PETER'S WOOD

o Bude

o Tintagel

PORTCOTHAN VALLEY

HAWKE'S WOOD +

o Wadebridge

B o d m i n

M o o r

Launceston
o

o Bodmin

RED MOOR +

PELYN WOODS +

LUCKETT +

St. Austell o

+ PELYNE

o Liskeard

Fowey
o

TAMAR ESTUARY +

Saltash
o

o Looe

0 1 2 3 4 5   10 miles

The Nature of Cornwall 1982
has been published in a Limited
Edition of which this is

Number 703

A complete list of original
subscribers is printed at
the back of the book

THE NATURE OF CORNWALL

THE NATURE OF CORNWALL

THE WILDLIFE AND ECOLOGY
OF THE COUNTY

BY

RENNIE BERE

ILLUSTRATED WITH DRAWINGS BY

MARJORIE BLAMEY
&
FRANKLIN COOMBS

WITH PHOTOGRAPHS BY
B. & S. BOTTOMLEY, COLIN BUTLER & JOHN BESWICK
AND MAPS BY V.S. PATON

IN ASSOCIATION WITH THE EDITORIAL COMMITTEE OF
THE CORNWALL NATURALISTS' TRUST
(Philip Blamey, Colin Butler & Kenneth Williams)
FOREWORD BY HRH THE PRINCE CHARLES
Patron of the Royal Society for Nature Conservation

BARRACUDA BOOKS LIMITED
BUCKINGHAM, ENGLAND
MCMLXXXII

NATURE OF BRITAIN SERIES

PUBLISHED BY BARRACUDA BOOKS LIMITED
AND PRINTED AND BOUND BY
NENE LITHO & WOOLNOUGH BOOKBINDING
WELLINGBOROUGH. ENGLAND

JACKET PRINTED BY
CHENEY & SONS LIMITED
BANBURY. OXON

LITHOGRAPHY BY
CHAPMAN BROTHERS LIMITED
KIDLINGTON, OXON

DISPLAY SET IN BASKERVILLE AND
TEXT SET PRINCIPALLY IN 10½/12pt BASKERVILLE
BY BEDFORDSHIRE GRAPHICS LIMITED
BEDFORD. ENGLAND

ISBN 0 86023 163 1

Contents

Nature is infinitely diversified, and yet each production makes its appearance at the time, and under the circumstances, which we would be led to expect. A plan which is so perfect and so harmonious, of which the parts are so diversified, and yet which naturally promote the existence of each other — which blend the sea, the land, the air, into one whole.

The only sure way to become naturalists, in the most pleasing sense of the term, is to observe the habits of the plants and animals that we see around us, not so much with a view to finding out what is uncommon as of being well acquainted with that which is of everday occurrence.

The long roll of the Atlantic upon the Cornish coast.

Robert Mudie in *The British Naturalist* — 1830

(Whittaker, Treacher & Co. London)

Foreword

By HRH The Prince Charles, Duke of Cornwall, Patron of The Royal Society for Nature Conservation

As Patron of the Royal Society for Nature Conservation I am delighted to be able to contribute something to this book. I suspect that Mr Blamey asked me to write the foreword as a result of a meeting we had last year in a plantation belonging to the Duchy of Cornwall. As a result of that meeting the Duchy has agreed to undertake various measures in the wood which will, we hope, contribute towards the survival of the Heath Fritillary Butterfly in Cornwall.

Before reading *The Nature of Cornwall* I would have said it was impossible to deal within the covers of a single book with so many different aspects of the wildlife of what is, in any view, a county with a vast diversity of habitats. As anyone who does read this book will discover, the authors have triumphantly solved the problem of setting out their material, in a way which not only makes the book immensely enjoyable for the armchair reader, but also ensures that more active readers who visit particular parts of the Duchy, or areas with a particular habitat, can find all the relevant information within the compass of a few pages.

The illustrations complement the text superbly and *The Nature of Cornwall* will be both a magnificent introduction to the wildlife area for those who have not yet been lucky enough to visit Cornwall, and an invaluable work of reference for all those who value, and wish to conserve, some of the most beautiful countryside in Britain.

Charles.

Introduction

This book is not about the myths, romance and rather fanciful view of history that many seem to look upon as being the very 'nature' of Cornwall. Nor does it purport to be a comprehensive account of the ecology and natural history of the county; such a work would fill many volumes. It is about the wildlife — a term which covers all the natural forms of plant and animal life — and is written for the ordinary reader who enjoys the countryside, the moors and the coast, and wants to know more about them.

The book is based on habitats. Following a background chapter which includes a very simplified account of the geology, there is a general description of the chief terrestrial habitats of mainland Cornwall and what lives in them. Then there are four chapters on the habitats most affected by human usage — some actually created by man. The later chapters describe briefly the important seashore habitats, the Isles of Scilly and finally the wider issues of conservation and the ways in which they are being tackled.

Technical jargon has been avoided as far as possible. It had been intended to omit scientific names altogether except in relation to species which have no common name: most lichens and mosses, for example. In the event this has not proved practicable. The English names of mammals, birds, fish, reptiles, amphibians, butterflies and moths are generally too familiar for there to be any possibility of error. This is not the position with invertebrates other than Lepidoptera; nor with plants, particularly the less common species. Even quite familiar plants are known by a variety of names. Wall pennywort, for example, is sometimes called navelwort, and one of the commonest plants in Cornwall, *Allium ursinum*, appears in print as wild garlic, wood garlic, broad-leaved garlic and ramsoms almost indiscriminately. In these groups, therefore, scientific names of the species — and only of the species — are included in the Index for the benefit of readers accustomed to this form of nomenclature. The text of this book was completed before publication of *A Review of the Cornish Flora* by L. J. Margetts and R. W. David (1981) so that plant nomenclature could not be reconciled.

Detailed information on the various species referred to can be found in the numerous Field Guides and other works of reference which are listed in the bibliography. This includes several books about specific aspects of Cornwall's natural history, four of which have been in constant use during the preparation of the present book: R. D. Penhallurick's outstanding studies of the avifauna, *Birds of the Cornish Coast* (1969) and *The Birds of Cornwall and the Isles of Scilly* (1978), Jean Paton's *Wildflowers in Cornwall* (1968) and Stella Turk's *Sea Shore Life in Cornwall* (1970).

Outside the granite areas and the unique formations of the Lizard peninsula, the soils of Cornwall are mostly derived either from the carboniferous culm measures of north-east Cornwall or from the Devonian slates, shales and sandstones of the rest of the county. To simplify differentiation between the rocks and soils of these main formations, the term 'killas' is here used only for the latter; the carboniferous culm measures are generally referred to as 'culm'.

Throughout the text the expression 'the Trust' refers to the Cornwall Naturalists' Trust. Only when there could be confusion with the National Trust (always referred to as such) has the full title been used.

The editorial committee — as individuals and not only as committee members — have assisted the author in many ways, particularly in regard to their own specialist interests. The work of the artists speaks for itself. It has been a privilege to work with them, and not only on this book, for both are active members of the Trust Council. Photographs are acknowledged individually. Most of those by Colin Butler and John Beswick were taken specially for *The Nature of Cornwall*, and this often meant travelling long distances in search of subjects in singularly discouraging weather. The Nature Conservancy Council and the outstanding team of Brian and Sheila Bottomley have been exceptionally generous with the loan of photographs, as has The National Trust.

The whole Trust is honoured by the Foreword kindly written by HRH Prince Charles, who takes a considerable personal interest in the conservation of nature in the Duchy.

Author and editorial committee would also like to thank the following for their help and co-operation without which the book could never have been produced: Frank Ansell; Mary Burgess; Harry Calder, the County Planning Officer; Dr Paul Chanin, Giles Clotworthy of the National Trust; Bob Darke; Rachael, Stephen and Tim Dingle (particularly in relation to farming practices); Julia Drage and the Information Department of English China Clays Ltd.; P.J. Dwyer; Dr Lewis Frost; Barbara Garratt; Russell Gomm and Pat Sargent of the Nature Conservancy Council; Allan Griffiths; Dr K.S. Hocking; George Jackson; Andrew Jewell; Mary Johnson; Avril Longman; Dr Gillian Matthews; Len Margetts; Jean and Pat Paton; Roger Penhallurick; Dr Franklyn Perring of the Royal Society for Nature Conservation; Ruth Phillips; Caroline Rigby; Fred Shepherd; Barbara Sturdy; Stella Turk; Bryan Wilson (particularly in relation to Duchy of Cornwall Woodlands); and the staff of the Cornwall County Library. We are also most grateful to Jackie Adams who typed the manuscript; Winwood Reade who read and criticized the text and gave much good advice; and Anne Irons who has carried out the formidable task of preparing the index. Clive Birch has been a most helpful and co-operative publisher.

All this notwithstanding, the author accepts responsibility for any errors. Opinions expressed are his and do not necessarilly represent the official views of the Trust.

RENNIE BERE

Bude
Cornwall
March 1982

Large Blue butterfly and Chough — our most recent losses. (MB)

The Background

A narrow Cornish Valley —
East Port Holland. (FC)

Cornwall, a peninsula jutting into the Atlantic, is almost an island cut off from the rest of the country by the River Tamar. Including the Isles of Scilly, the county covers an area of 876,015 acres and is 86 miles long from Marsland Mouth to Land's End. No part is more than fifteen miles from the sea, most of it much less, the result being a mild climate — generally warmer in winter and cooler in summer than the rest of southern England — with above average rainfall and strong, often gale-force, salt-laden winds sweeping across a landscape marked by exciting contrasts. While exposure to wind and salt is uninviting to most forms of wildlife, this flourishes in unusual abundance in sheltered pockets where warm and wet conditions prevail.

The most obvious contrast is between the two coast lines. The north is harsh and wild; high, bare cliffs, little relief and few harbours. The south is gentler, the cliffs lower, less abrupt and in general with more vegetation. Most of Cornwall's rivers flow towards the south and open into harbours and estuaries; these break the line at frequent intervals and are bounded by woods and edged by salt-marsh where wading birds and wildfowl feed. But even the north coast provides its own contrasts. A change of angle and the rock face suddenly reveals cushions of brightly coloured flowers. The stunted oak forest of the Dizzard, which stretches down landslip slopes almost to sea-level, has alongside it some of the most uncompromising cliffs in Cornwall. There are also sand-dunes and stretches of blown sand. The Camel Estuary is the one major break in this coast, a haven for many forms of wildlife as it was for troubled shipping in the days of sail. Whether on the north coast or the south, the estuaries show the continuing contrasts of the tides: there may be little to see when the tide is in but, as it recedes, mud-flats are exposed, revealing unsuspected vegetation and birds arriving to probe for food. Indeed, wherever the sea meets the land, the scene is always changing according to the state of the tide. Every shore — sand, shingle or rocky platform — has its own plant and animal communities, often in well defined zones. At one end of the daily tide cycle, the land and its inhabitants are totally submerged. At the other, they are completely exposed to the wind, the sun and the rain, mussel beds and banks of seaweeds again on view.

Inland there are the cultivated farm lands with pockets of industrial dereliction of a peculiarly Cornish kind. There is also the contrast between the ferns and colourful flowers in deep-cut lanes, the mature old hardwoods in lush valleys and the apparent blankness and lack of variety on the moors. This blankness is more apparent than real, for there are glorious pockets of life in the wetlands, where orchids and cotton grass thrive, introducing yet another contrast. Many Cornish hedges (stone walls) are almost bare, except for gorse and lichens, on the windward side but full of colour on the other with butterflies resting in the sun and digger wasps making their nests.

Soils tend to be acid and to relate closely with the underlying metamorphised sedimentary and granitic rock. The former consists mainly of the carboniferous culm measures of north-east

Cornwall and the Devonian shales and sandstone grits known as 'killas'. These are the remains of the heavily folded and faulted Armorican mountains, which stretched down the south-west peninsula some 250 million years ago. Molten rock from the earth's interior welled up into the heart of these mountains. It cooled and consolidated under pressure and was gradually exposed by erosion, to become the granite moorland spine of Cornwall: Bodmin Moor, Hensbarrow, Carmenellis and West Penwith where the granite reaches the sea. Some twenty-five miles from Land's End it reappears from the Atlantic to form the archipelago of the Isles of Scilly, a group of 145 rocks and low-lying islands. Though the granite moors are of no great altitude — Brown Willy at 1,377 feet is the highest point — they are everywhere higher than the surrounding countryside.

Derivatives of granite form the background of Cornwall's main land-based industries — mining and extraction of china-clay. Hot gasses and fluids, which arose from the granite as it cooled, became mineral deposits of tin, copper and other metals. These have been mined in Cornwall for centuries — though the scope of operations has declined since the high point of the 19th century, there are current signs of revival and of interest in underground sources of energy in the form of a 'hot rocks project'. Industry has left a scatter of derelict engine-houses, old mine buildings and heaps of waste, in various stages of recolonization by plant life throughout a large part of the county.

China-clay, or kaolinized granite, has been dug for over 200 years. The huge piles and hillocks of waste (coarse quartz-sand and mica from the decomposed granite) have produced the familiar 'lunar landscape' of Cornwall: the most obstinate collection of 'holes and heaps in juxtaposition' to be found anywhere in the nation, according to John Barr, and a backlog of dereliction which leaders of the industry are now beginning to tackle. There are also quarries and abandoned quarries throughout the county, quarrying granite and other local rocks being among Cornwall's oldest industries. Though few are worked today, there are many disused quarries, particularly on the moors, where they support interesting ferns and other plants as well as providing nesting places for birds. The Delabole slate quarry, which has been worked for four centuries and is over 500 feet deep, still produces high quality slates.

As well as the killas, culm measures and granite which underlie most of Cornwall there are several small exposures of very hard greenstone: to the north and west of Land's End granite cliffs, Nare Head, Caragloose Point and one or two other places. There are also two zones of volcanic rock, one of which follows the line of the River Inny and continues on to Tintagel and Pentire Head, where the lavas display a beautiful pillow-like structure produced by cooling under the sea millions of years ago. The oldest exposures in Cornwall, however, are the ultra-basic (alkaline) pre-Cambrian rocks of the Lizard peninsula, whose plateau surfaces consist largely of red and green serpentine with intrusive masses of boulder-strewn gabbro, mica-schists and hornblende-schists. While the flat plateau is scenically rather uninteresting, the cliffs and coves are superb, and the soils produced by these ancient rocks support a unique flora, including lime-loving plants which are not frequent in Cornwall. Limestone is found only near Torpoint, though there is a band of cherts between Launceston and the north coast. Beaches and sand-dunes are rich in lime, but this is produced by shell fragments.

The harder strata among the sedimentary rocks have given rise, not only to some of the finest cliffs, but also to those moors and downs which are not part of the granite: St Breock Downs, Laneast and Wrasford Moor north of Kilkhampton, for example. Slate is a major constituent of the killas. It occurs on the south coast and in bands elsewhere. One of these underlies the southern fringe of Bodmin Moor, continuing on to the north coast to produce the spectacular cliff scenery around Bedruthan Steps. Where softer shales predominate, there are less stable slopes, and cliffs are virtually absent, as at Perranporth and Widemouth Bay.

The highest, though perhaps not architecturally the finest cliffs in Cornwall, are the sheer cliffs, built up of alternate strata of shales and sandstones where the culm measures are exposed to the Atlantic. High Cliff (731ft), Dizzard Point and Henna Cliff are all on this part of the coast where there are striking examples of folding, faulting and thrusting of the various layers in the culm, including the remarkable cascade of zigzag folds exposed at Millook. Elsewhere, the more resistant sandstones form buttresses and slabs of rock which stand almost vertically upright.

The level of the sea in relation to the land has varied greatly throughout geological history. At one time during the Tertiary Period almost the whole peninsula was submerged beneath the sea, from which it has emerged in slow uneven stages. This has produced the ancient marine platforms, or plateaus, which give so much of Cornwall a flat appearance and which can be seen at different places and levels from the '1,000 foot platform' on Bodmin Moor to the raised beaches of the immediate pre-glacial period, which are often apparent just above the high-tide level. Cornwall was not covered by ice during the great Ice Ages of the Pleistocene but higher ground must have been snow-bound, with the rest of the land an ice-margin type of tundra. As the climate warmed up, melting snow and heavy rains caused flooding, which gouged out valleys and estuaries and inundated forests of the coastal plain; submerged forests have been located at several places including Mount's Bay, St Austell Bay and Bude.

Since the end of the last Ice Age, about 10,000 years ago, further changes of climate have caused a degree of infilling in many estuaries as is shown by several old ports now well away from the sea: Lostwithiel and Hele Bridge, for example. Blown sand has not only created extensive sand-dunes (known as 'towans') but has also buried several buildings. Sand and gravel bars have been formed: among them the Doom bar across the Camel Estuary, the bar of Tresco in the Isles of Scilly and the shingle bar near Porthleven which has dammed the Cober river, created Loe Pool, and closed the sometime tidal port of Helston.

The tundra-type landscape of Cornwall was first occupied by scattered herds of reindeer and animals such as the Arctic fox. With improving climate the tundra was invaded by trees; animals which thrive in open country were replaced by woodland species such as deer, wolves and wild boar. The early hunter-gatherers gave way to Neolithic peoples from the continent, who cleared some of the wildwood in order to cultivate the land — the first move in man's assault upon the environment. They grew corn and kept flocks of sheep, goats and cattle but continued to hunt for some of their needs. The mammal fauna must have been much as it is today, except that elk, aurochs and beaver were present, and rabbits were absent until the Normans introduced them to Britain. There were also mammals which have recently become extinct locally, (polecat and pine marten, for instance), or which survive precariously like the otter, once plentiful in rivers and around the coast.

These are examples of the sort of changes which have always been taking place in the natural world — and not only in Cornwall. The pattern of change is now accelerating. It is happening in relation to woodlands, farming practice, river management, the drainage of wetlands and chemical pollution of the air, the land and the oceans; the effect on wildlife has usually been adverse. In his *Birds of Cornwall,* for instance, Roger Penhallurick lists thirteen species recently extinct in Cornwall as breeding birds and a further fifteen as now being 'exceptional breeders'. These could soon be joined by several familiar species such as puffin and particularly nightjar. Gannets once bred on Gulland Rock off Trevose Head (first recorded by William of Worcester in the fifteenth century) and on the Isles of Scilly but those now seen off the coast breed elsewhere. Common sandpiper, Sandwich tern, snipe, redshank and dunlin have nested in Cornwall, mostly in small numbers, but do so no longer; nor do the long-eared or short-eared owls which bred regularly until the 1940s. The Cornish chough finally became extinct in the late 1960s. Redpoll and collared dove, by contrast, have recently started breeding.

The relentless decline of Britain's birds of prey has been well publicised and still continues. Both osprey, now quite often seen as a winter visitor, and red kite once bred in Cornwall as did marsh and Montagu's harriers, the latter regularly in small numbers until 1973. As recently as 1955, there were seventeen occupied peregrine eyries; now only two or three pairs of this magnificent raptor rear their eyasses on the Cornish coast. Kestrel and buzzard, the 'rabbit hawk' of old Cornwall, seem to be becoming steadily fewer on the north coast with every passing year. Many kinds of sea-bird appear doomed as a result of oil pollution; and many common land birds may be in danger of suffering a similar fate because of chemical pollution and the widespread destruction of habitats. Meanwhile the populations of those species most capable of adapting to the ways of man are gradually expanding: herring gull, starling, wood pigeon and the more robust corvids, notably jackdaws. Woodland birds survive where deciduous woodlands survive and in some of the more modern mixed plantations. Gardens, fortunately, still provide nesting and feeding places for a variety of species, but only where gardeners show restraint in the use of chemical sprays and control their passion for excessive tidiness.

The large blue butterfly, emblem of the Cornwall Naturalists' Trust, became extinct in Cornwall in 1973 and throughout Britain shortly afterwards; desperate efforts are being made to prevent the heath fritillary following suit. The red ant *Myrmica sabuleti*, with which the blue associates, is also on the decline. Of the eighteen species of ant recorded in Cornwall, seven have not been reported since 1950. Bumble-bees tend to be linked with certain flowers. If the plant population declines so does the bee. *Bombus jonellus*, for example, collects most of its food from the heath family, and its distribution is dictated largely by the abundance or otherwise of heathers.

Barren Strawberry, Cornwall's first recorded plant. (MB)

The earliest known published reference to a plant growing in Cornwall (the barren strawberry) was made by Mathias de Lobel, after whom the lobelia was named, in 1576. In the *Survey of Cornwall* (1602) Richard Carew mentions various plants and from this we know that many still familiar species flourished 400 years ago: among them common and western gorse, sundews, sea holly (now becoming scarce), thyme and wood garlic, 'the countryman's treacle' according to Carew; he only mentioned three kinds of tree: Spanish chestnut, hazel and 'oke'. Since then numerous accounts of the wild flowers of Cornwall have been published culminating in F. Hamilton Davey's *Flora of Cornwall* (1909) updated in 1981 by L.J. Margetts and R.W. David's *A Review of the Cornish Flora*. Davey listed seventeen species, including six from the Lizard, as not being recorded in any other English county; among the seventeen were western fumitory, a maritime scabious and Cornish heath. He also mentioned species which had recently become extinct in Cornwall: lesser fleabane, which disappeared when some rough ground was levelled near Drift, and hoary cinquefoil, rooted out by a collector fifteen years after its discovery.

Endangered species. (MB)

Bog orchid

Heath lobelia

Spotted catsear

Flax–leaved St John's Wort

17

Since then the little sand crocus, sea clover, strapwort, four-leaved allseed and the prostrate purple spurge have been lost; several other plants are in serious danger of following them. Among these are the heath lobelia, bog orchid, spotted catsear, Watling Street thistle and the flax-leaved St John's wort. Meanwhile exotics are naturalising themselves and may eventually come to be accepted as part of the flora of the county. Among them are the evergreen holm oak, three-cornered leek, several varieties of narcissus, *Erigeron glaucus* with daisy-like flowers, periwinkles, the colourful wild mesembryanthemum from southern Africa and a Spanish tree-heath naturalised on railway cuttings near Liskeard.

The recent loss of certain plants and animals and the introduction of others emphasise the importance of systematic recording, as changes take place in the countryside. Recording not only species but sites of interest to naturalists has always been looked upon as an important function of County Trusts.

Torrey Canyon — a disaster. (FC)

ABOVE: High bare 300 ft cliffs and vertical stratification on the north coast
culm. (JB) BELOW: Millook cliffs — zigzag folding and thrusting of the
culm strata. (JB)

OPPOSITE ABOVE LEFT: Portquin Bay — coast and cliffs east of Pentire; (CGB) RIGHT: Whale's Back Rock, Bude — a periclinal arch in the sandstone; (JB) CENTRE: St Austell Bay — the more gentle south coast (CGB) and BELOW: a tidal south coast estuary — Tresillian river at St Clement. (CGB)
ABOVE: Open ground on Bodmin Moor with sheep scratching on an ancient hut circle; (JB) BELOW: The upper rocks of Rough Tor. (JB)

ABOVE: The Isles of Scilly are an extension of the Cornish granite; (NCC)
BELOW: Roche Rock at the edge of the St Austell granite — tourmanilised
granite rising above the killas. (CGB)

ABOVE: The search for underground sources of energy — the geothermal
project at Longdowns near Falmouth; (CGB) RIGHT: Nesting Buzzard —
the 'rabbit hawk' of old Cornwall, (JB & SB) and LEFT: oiled guillemot — a
victim of marine pollution. (JB) OVER: Mullion Island. (NCC)

Coast and Cliffs

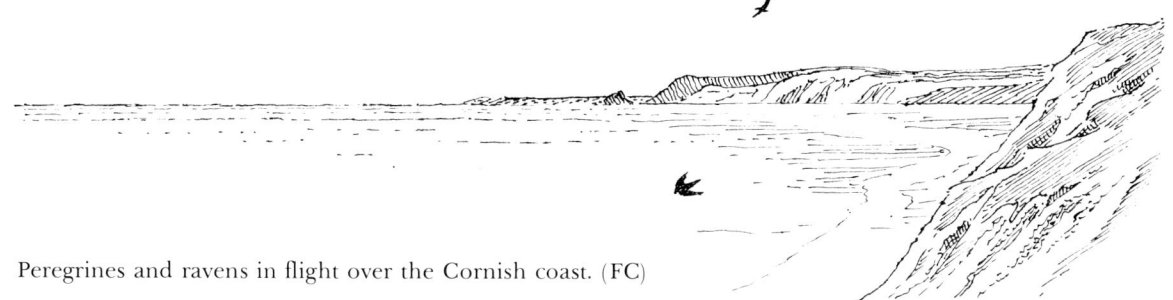

Peregrines and ravens in flight over the Cornish coast. (FC)

The term 'coast' covers a number of distinct habitats — cliffs and cliff-top land, coastal valleys, rocks, beaches, foreshore, dunes and estuaries. Together, marine pollution notwithstanding, they are the habitats least altered by man. This chapter is concerned with the coast land above the high-water mark. Beaches, rocks and estuaries directly influenced by the sea are looked at in later chapters. Unless you are delving into these, the best way to see the coast of Cornwall is on foot, by the coast-path, which can be taken in short sections by those who prefer — it is indicated on the map of Cornwall inside the front cover of this book. There are also many points from which great sweeps of the coast-line may be seen: Efford Beacon, Kelsey Head and Dodman Point particularly. Except for a small and rather untypical area in Penwith there are no Trust reserves actually on the coast. Many long stretches, however, are owned by the National Trust and are thus safeguarded in perpetuity. Nature in Cornwall has benefitted greatly from purchases made under the long-sighted Operation Neptune.

The magnificent cliffs of north Cornwall form the edge of an elevated plateau, which is cut by short and deeply incised valleys, some of which are truncated to produce waterfalls to the shore as at Litter Mouth, Tidna Valley and the Dizzard. By contrast Marsland, Crackington and Valency valleys are cut right down to sea-level. Writing of this area in *Westward Ho!* in 1855, Charles Kingsley described each valley as opening 'through its gorge of down and rock towards the boundless ocean. Each has its upright walls, inland of rich oakwood and nearer the sea of dark green furze, then of smooth turf, then of weird black cliffs which range far out into the deep sea'. The oakwoods are still present in many of these valleys, but much of Kingsley's smooth turf is now arable or — since myxomatosis decimated the rabbit population — scrub.

One result is that the large blue butterfly, which had its last Cornish home on this coast, became extinct in the county in 1973 and throughout Britain six years later. It has a remarkable life-cycle depending entirely upon wild thyme, and the red ant *Myrmica sabuleti*. The large blue's eggs are laid on thyme where they hatch after 7-10 days. After three weeks, the caterpillar falls to the ground and starts wandering. It meets the ant which is attracted by an excretion from the caterpillar's gland. The ant then takes the caterpillar to its nest where, in its turn, the caterpillar feeds on the ant's larvae and pupae. Hibernation and pupation follow; the new butterfly emerges, having spent some ten months in the ant's nest. Both ant and thyme thrive in closely cropped grassland and thus are vulnerable to changes in farming practice and in the rabbit population. Efforts to save the large blue were made throughout the 1960s and 1970s, the main emphasis being placed upon thyme. It was not realised until too late that a high population and density of the ant, distributed over a wide area of very short turf were the essential prerequisites of the butterfly's survival.

Many other butterflies are still to be seen in the coastal valleys where the large blue used to thrive: among them are the common blue, grayling, marbled white, small pearl-bordered and other less common fritillaries. Both brown and green hairstreaks are occasionally present: the former usually near hawthorn scrub, the latter where gorse and grassland plants flourish. The white-letter hairstreak, which is uncommon in Cornwall, has been observed on a thistle-head in one of these valleys. Among other insects the great green bush-cricket and the common field grasshopper are worth looking for. Tawny cockroaches are quite common in low herbage on cliff tops and the large dark bloody-nose beetle crawls about slowly in the grass, feeds on bedstraws, and emits a red fluid from its mouth when handled.

Between Marsland Mouth and the Camel estuary there is a whole series of exciting incidents with high points at Henna Cliff (470 feet), Beeny (469 feet) and High Cliff (731 feet). There are vertical cliff strata in the culm, zigzag folds of shales and sandstones exposed to optimum effect at Millook and the Bude 'fish-beds' with dark sandstone nodules which enclose remains of the fossil-fish *Cornuboniscus budensis*. West of Boscastle, where the culm gives way to the killas, the cliffs are among the finest and most varied in Cornwall, with coves, headlands, off-shore stacks and islands, particularly around Tintagel. There are volcanic rocks and slate outcrops between Portgaverne and Portquin, the luminous moss which reflects light from crevices at Lundy Bay, the greenstone of the Rumps and the pillow lava cliffs of Pentire. The only worthwhile breaks in the cliffs are at Widemouth Bay and Bude where the relic sand-dunes still support a few plants of sea holly, sea rocket, eastern rocket and wild leek.

Beyond Stepper Point, the western portal of the Camel estuary, there is a fascinating stretch of coast with strange blow-holes created by erosion (Pepper Hole and Butter Hole), which link the cliff-tops with the sea below; behind Harlyn Bay is a low-lying marsh which brings reed-beds and wetland birds to the edge of the Atlantic. The coast changes in both direction and character at Dinas Head. There are the remarkable stacks and rock formations of Bedruthan, formidable headlands with their wealth of coastal flowers at Cligga Head, St Agnes Head (with the Beacon, 629 feet, a short distance inland) and Godrevy, and the colourful heathland of Carvannel and Reskajeage Downs with bell heather and western gorse —'the rugged seaboard of industrial Cornwall'. There are also the blown sand and dune areas of Constantine Bay, where Italian cornsalad grows, and the half-buried church bears witness to the mobility of sand, the much more extensive Perranporth dunes and the towans (Cornish for sand-dunes) of the Hayle-Gwithian area beyond Godrevy.

Spring squill

Flowers of the cliff tops. (MB)

Gorse and dodder.

The wild plants of the coast are abundant throughout, the wooded valleys adding to the interest. These are mostly dominated by oak with sycamore, ash, holly, hazel and hawthorn, with colourful flowers and tall grasses wherever there is a break in the canopy. Few of the woods have been subjected to more than a minimum of management, though some are now being replanted, mainly with conifers. It is the cliffs and cliff-tops, however, which are particularly attractive. In early summer the exposed edges of maritime grassland are outlined with colourful thrift; gorse often garlanded with parasitic dodder, self-heal, thistles, sheep's sorrel, buttercups, spring squills and eyebright colour the grassland. Between bands of scurvy grass, the cliff faces are painted yellow by bird's-foot-trefoil and kidney vetch, a plant which shows considerable colour variations on this coast. Rock samphire, cliff spurrey, sea campion and wild carrot grow where patches of turf allow them to take root, with — closer to the sea where waves beat against the rocky lower slopes — rock sea-lavender, lichens and algae. Where stone walls and Cornish hedges are exposed to the weather there are stonecrops, cliff spurrey, wall pennywort and several ferns — hart's tongue, black and maidenhair spleenwort are the most common. Sea spleenwort and wild maidenhair ferns grow in remote rock crevices.

Throughout this long coastline birds are just as interesting as flowers. Fulmars breed at intervals all along it from the far west to Henna Cliff in the north, as do shags — but few cormorants — and gulls, with Cornwall's main kittiwake concentration on St Agnes Head. No auks breed north of Beeny, near Boscastle, but they are plentiful south and west of that point: guillemots the most numerous, then razorbills and puffins. These outstandingly attractive birds, with their narrow wings and fast whirring flight, are the principal avian victims of the modern curse of oil pollution. Ravens still breed on the cliffs, and there are a few breeding pairs of peregrines, whose favourite victims are the rock doves and feral pigeons which are so plentiful. Kestrels and buzzards are perhaps less often seen than they used to be. Manx shearwaters pass along the coast on migration and are frequently forced ashore by severe weather. Oystercatchers are to be seen along the beaches throughout much of the year, but breeding sites are mostly north of the Camel. Rock-pipits stay below the cliffs; meadow pipits are among the most plentiful birds of the coastal grassland, while stonechats and linnets prefer the furze. Jackdaws — but no longer choughs — are numerous everywhere. A few wheatears nest but most observations are of passage migrants.

A number of badgers and foxes live in the coastal scrub. Rabbit populations seem to be increasing, and a few grey squirrels have adapted to the cliffs where there are no trees. The commonest small mammals are the short-tailed vole and both common and pigmy shrews. There are stoats and weasels, and moles in tunnels wherever there is suitable soil. Natterer's, Daubenton's and Barbastelle bats have all been collected within a few yards of a beach in north-east Cornwall, but their roosts were almost certainly in a small wood, not actually part of the coast. Reliable information about bats in the county is distinctly patchy. These unusually interesting animals deserve more attention from naturalists than they usually get. A detailed census of grey seals has not been attempted for many years, but numbers may be declining slightly, though all the traditional breeding caves are still occupied. The largest colony on the coast of mainland Cornwall is just north of Boscastle.

There is so much natural abundance, that it seems invidious to select a few special sites, but perhaps the most unusual is the stunted oak forest of the Dizzard which stretches down landslip slopes for nearly 400 feet almost to sea-level. Exposure has resulted in a tight wind-pruned canopy with trees only five or six feet high, except in a few sheltered pockets. The trees are mostly oak with birch, rowan and one or two wild service trees, which are rare in south-west England. Beneath this strange canopy and the similarly stunted understorey, 120 flowering plants have been identified including wood vetch, wood garlic, sheep's sorrel, sanicle, heath pearlwort, small bristle club-rush, bluebell, wild strawberry, primrose and dog's mercury. Damp areas are dominated by

Familiar flowers of the cliff face. (MB)

Scurvygrass, Birdsfoot trefoil, Kidney vetch.

Rock samphire, Sea campion, Cliff spurrey

28

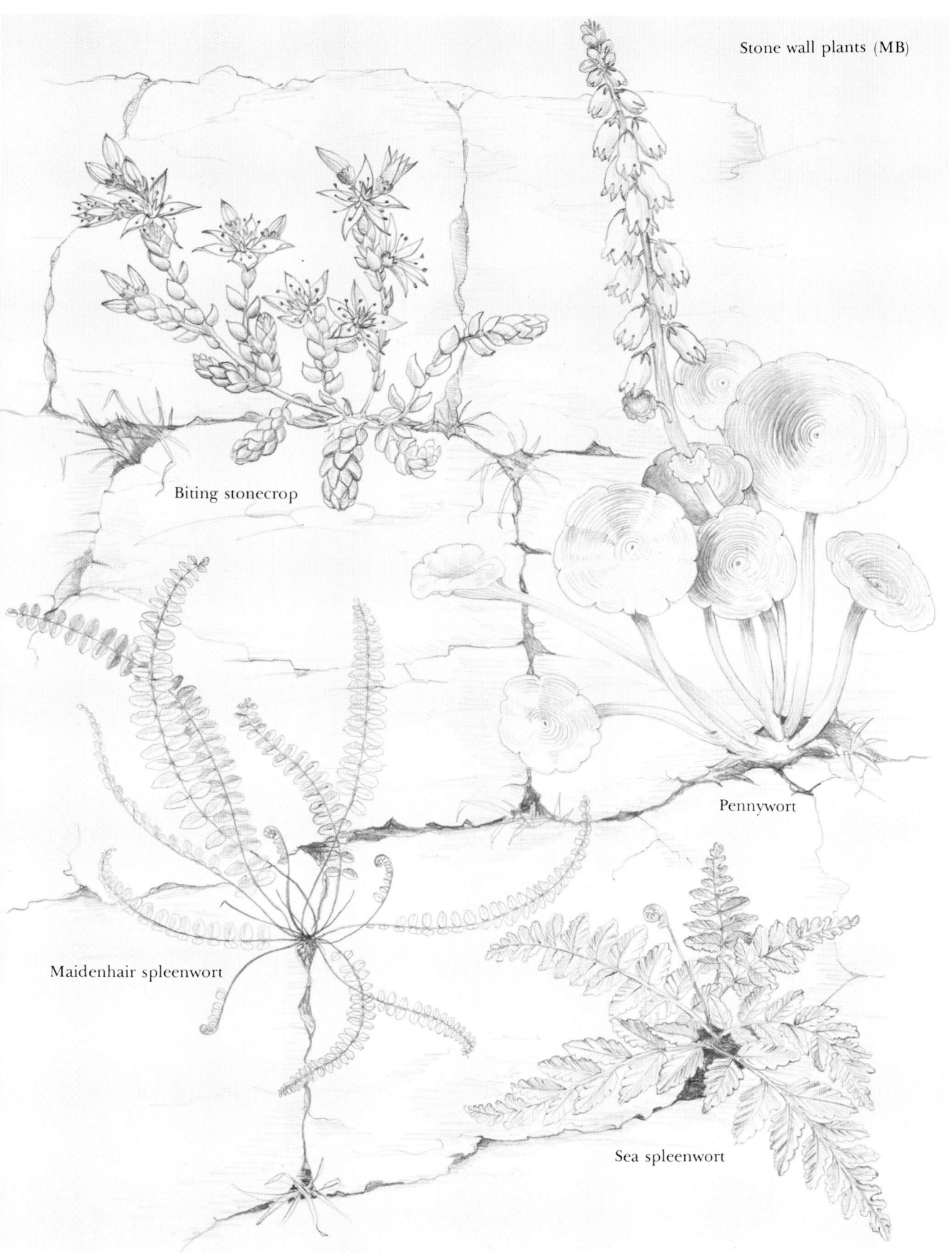

Biting stonecrop

Pennywort

Maidenhair spleenwort

Sea spleenwort

bilberry, cow-wheat and tufted hair-grass. Ferns, including broad buckler and royal ferns, are abundant. Many of the trees are swathed in moss and decorated by lichens, notably the blue-grey *Stricta limbata* and the yellow-orange *Pseudocyphellaria crocata,* which is also found at Bocconoc but nowhere else in England or Wales. The large plate-shaped tree lungwort grows on oaks throughout this surrealist forest where — as if to emphasise its unusual qualities — there once was a small heronry.

Dizzard flora (MB)

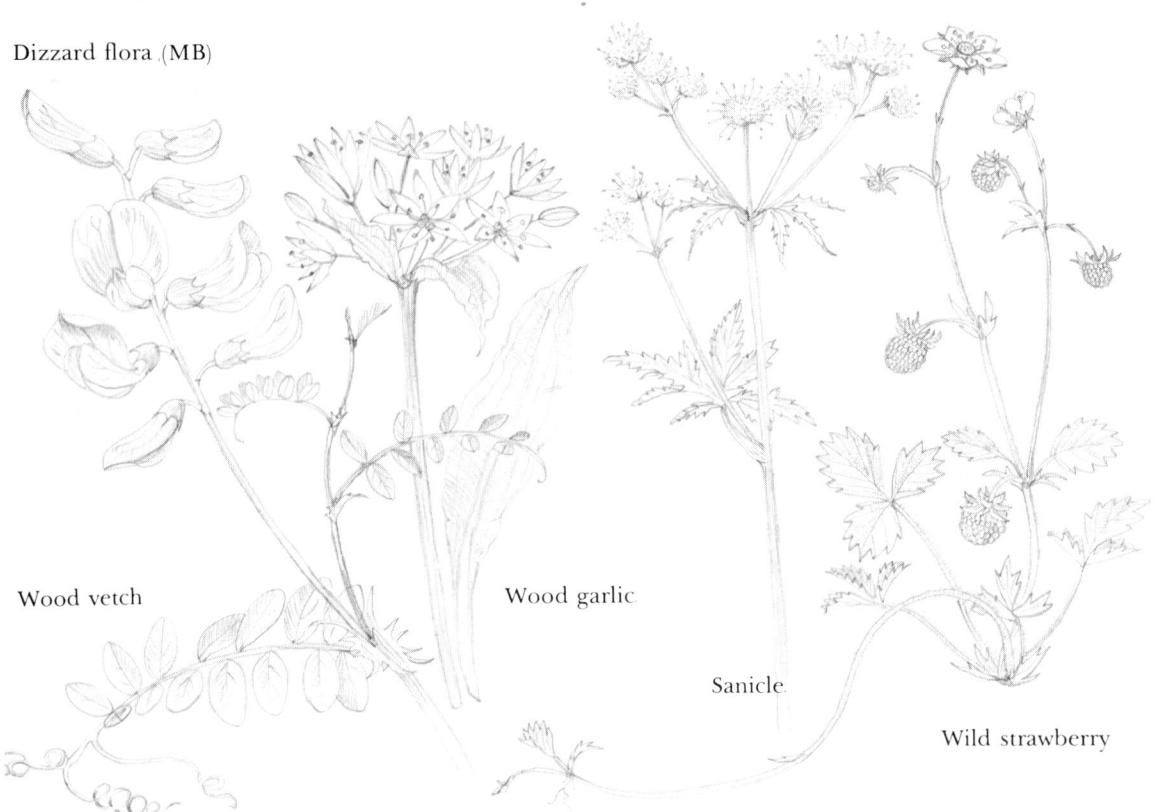

Wood vetch

Wood garlic

Sanicle

Wild strawberry

On Forrabury Common, at Boscastle, there remains what is probably the only British survival of the ancient Saxon (not Cornish) 'stitch-meal' system of land tenure, with some forty strips cultivated individually between March and November, and thrown open to common grazing for the rest of the year. For some years the Trust held an uncultivated strip as a reserve chiefly because the rare Fyfield pea grows upon it; the reserve had to be abandoned owing to management difficulties. Snails are particularly numerous on the common, probably because of the underlying rock strata containing calcium, required by these animals for their shells.

The Pentire peninsula is composed mainly of slate, with the intrusions of pillow lava which form Pentire Head itself and the high cliffs leading to the greenstone promontory of Rumps Point. Vegetation varies, with north-facing slopes largely covered by scrub, those facing west being maritime sward, with the usual coastal plants including spring and the much less common autumn squills abundant in season: prostrate Dyer's greenweed and the familiar cowslip which has a restricted distribution in Cornwall, are also present. Minver Hill, above Pentireglaze Haven, is a particularly good site with least and lesser bird's-foot-trefoil, small-flowered buttercup and the handsome musk-thistle, which has fragrant, drooping flower-heads. Nearby, below low cliffs, is a raised beach with sedges, rushes and reeds in damp patches. There are at least three badger setts on Pentire, one deep in a garden of bluebells. Seals breed below the Rumps and birds abound: seabirds as well as the usual cliff-top species such as stonechat, linnet, meadow pipit and jackdaw.

Dinas Head, beyond Trevose, is a complex geological formation with slate, greenstone and lavas. Many flowering plants flourish in the flat cliff-top grassland with an abundance of squills in spring but none in autumn. Depressions have produced wet flushes with various wetland plants, including bog pimpernel more usually found on moorland. There is some bell heather. Tree mallows line the steep slopes of Round Hole, which plunges down to the level of the sea. The base of the cliffs is almost devoid of vegetation but the upper levels are covered with thrift, rock samphire, kidney vetch and buckshorn plantain. Nearer to the headland golden samphire, rock sea-lavender, which is nowhere common in Cornwall, and prostrate wild asparagus occur.

Dinas Head (MB)

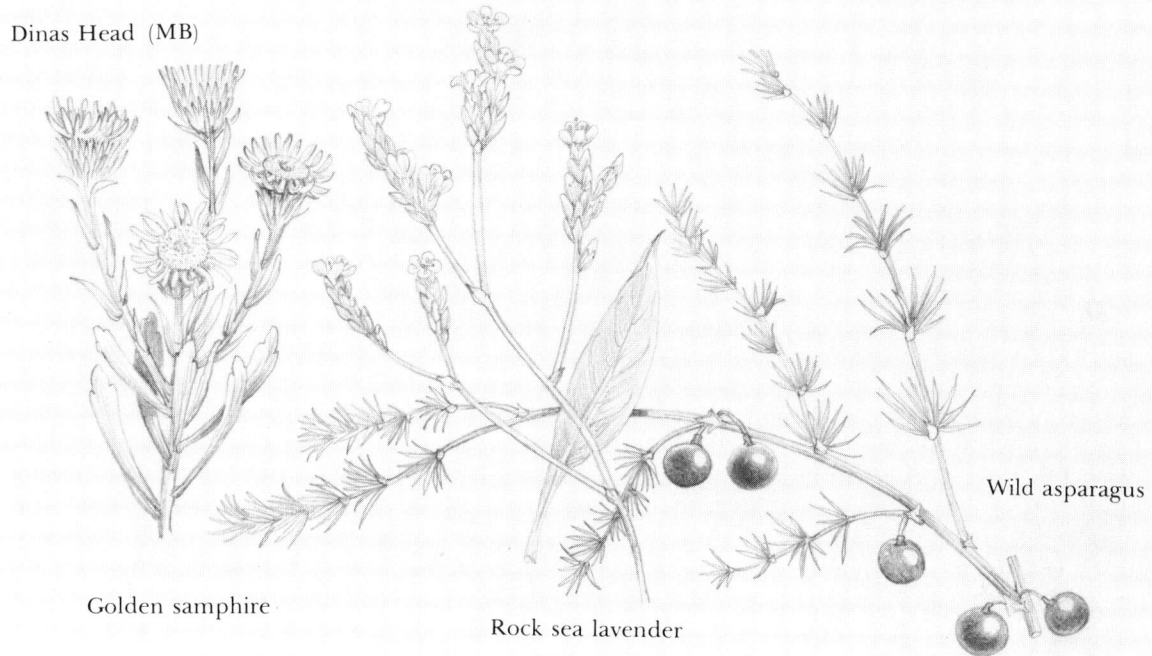

Golden samphire

Rock sea lavender

Wild asparagus

The Constantine Bay dunes are less extensive than the towans further south but support many of the same plants. They are highly mobile and lacking in 'slacks' but damp ground beside a stream, in which horned pondweed grows, brings to the area such wetland plants as yellow flax. Most of the more stable dunes are occupied by a golf course but there is room for flowers in the rough: among them the pyramidal orchid and 'devil's saffron', as Davey called the sand-dune form of dodder with yellow or orange stems. A short distance south and about a mile inland is the Trust's Portcothan reserve: a lightly wooded valley with a stream and an old quarry. It is mainly of interest because of its varied bird-life; some 75 species have been recorded in an area of only sixteen acres with breeding wheatears, buzzards and several warblers, with heron and curlew among regular visitors.

Beyond Watergate Bay, Newquay and the Gannel estuary are the Kelseys: Porthjoke and Holywell Bay with Kelsey Head between. There is a good variety of maritime plants along the cliffs and cliff-top grassland and in an area of fixed dunes. Seabirds are plentiful, and a few seals breed below the headland. Except for Penhale sands, however, the best botanical area on this stretch of coast is between the cliffs of Cligga Head (300 feet) and nearby Pen-a-gader. The cliffs and underlying rocks are granite and killas grit; the area is rich in minerals and the abandoned mine-workings which are the result of this. Ling heather is the dominant species, as is usual on re-colonised mine-spoil, and there are patches of willow carr near the site of a wartime airfield. Around Pen-a-gader, where the cliffs face north-west and exposure is extreme, there is some splendid maritime heathland dominated by western gorse and bell heather which often hides the woolly-leaved mountain everlasting. There are spring squills and eyebright, both normally grassland as opposed to heathland plants, and the rare hairy greenweed.

Except around Porthtowan and Portreath, there are few sand beaches between St Agnes Head and Godrevy. For the most part the sandstone cliffs drop sheer into the sea with 'abrupt magnificence'. The cliff-top land is mainly flat with, north of Portreath, abundant mining debris in various stages of re-colonisation by ling heather. Immediately south of St Agnes Head is a great unbroken bank of English stonecrop over 200 yards long. The vegetation is mostly maritime heath, gloriously colourful in season, but it is more varied near Porthtowan, where lime-rich, blown sand covers much of the topsoil and the rare spotted catsear grows. There is grassland at the top of the cliffs with kidney vetch demonstrating colour variations which are unusual even for Cornwall. The flora of the lower slopes resembles that of the distant chalk-downs with wild sage, ploughman's spikenard and lesser chickweed. Carvannel Downs, south of Portreath, are cut by deep valleys with small waterfalls to the shore; sea spleenwort occurs, and there is a fine display of thrift.

The rocks of Godrevy Head are alternating slates and sandstone with a Pleistocene raised pebble beach below the low western cliffs. The turf above is much influenced by blown sand from the towans and is covered in late summer by an abundance of autumn squills. Towards Navax Point, at the eastern end of the headland, heath takes over from grassland. Below are two seal breeding caves. One of these, between the point and Reskajeage cliffs, is accessible from the land only by way of a precarious old mine adit which penetrates the roof of the cave. It was the scene of the hair-raising seal-hunt by torch-light described by J.C. Tregarthen in *Wildlife at the Land's End*.

The two major dune areas, near Perranporth and Hayle, are believed to date from the 11th century AD and to have been built up on poor grazing land by wave action and wind. Because it consists largely of shells, the sand is extremely calcareous (calcium carbonate content about 60%), the result being a flora with many lime-loving species which are not common in Cornwall. The dunes themselves show typical stages in the succession from bare sand to stable dunes well colonised by plant life; this usually begins with marram-grass or sand couch-grass on the fore-dunes. Damp slacks at all levels support their own flora with such plants as sedges and yellow flag. The build-up of sand did not happen all at once but seems to have been the result of periodic 'blows', when the sand was unusually mobile because of high wind or surface disturbance. Buildings were buried at such periods — Upton Barton, for example, vanished from sight about 1650 and now lies some twenty feet below the surface. Over the centuries, a great deal of sand has been removed for use as a lime-dressing on agricultural land and, more recently, by broccoli growers. Holiday camps were established before the days of proper planning controls and the resulting increase in human use has greatly reduced the plant-cover.

The Perranporth dunes — at 270 feet the highest in Britain — are about three miles long and a mile and a half wide. Both the southern part of the area, where there is a holiday camp, and the seaward fringe have suffered greatly from human disturbance, but inland there is a worthwhile flora. The northern dunes, known as Penhale sands, have been occupied for many years by the Ministry of Defence; access has been restricted, and this has greatly benefitted the vegetation. Just above the high-tide line there are plants such as sand couch-grass, sea rocket, sea sandwort, orache and sea holly. Immediately behind this area, the steep dunes are too mobile to be colonised but higher slopes above are more gentle and support spurges, sea bindweed, lady's bedstraw, wild thyme and sand sedge. Further inland kidney vetch, cinquefoils and creeping thistle grow among the grasses. On the upper plateau there are carline thistle, docks, primrose, weld and groundsel with isolated thickets of hawthorn, privet and elder; and finally there is short turf, with typical grassland plants, maintained by rabbits.

The Hayle-Gwithian complex encloses much of St Ives Bay, but the dunes have been greatly altered by human use. Godrevy Towans, at the base of the headland, is a fixed dune area cut off

Sea rocket

Sea sandwort

Sea holly

Wild thyme

Sea bindweed

(MB)

from the rest of the complex by the seriously polluted Red River, which has covered much of the sand in its neighbourhood with mineral deposits brought down from the mines. There still are a few worthwhile plants on what is left of the dunes at Gwithian, but much of the best part of the whole area is Upton Towans; for many years human access was forbidden, initially because of an explosives factory, and more recently by ICI. Though no longer closed, human access is still slight and, as a result, the vegetation is similar to that of Penhale sands; additionally there are the pyramidal orchid, biting stonecrop or wall-pepper and, growing among the scrub in protected hollows, balm-leaved figwort, which is rare away from these towans. The rest of the area can best be described as a calamity. The once fine Phillack Towans have been partly cleared of sand and are largely occupied by chalet camps. Human trampling has destroyed most of the vegetation between the camps, and 'garden escapes' have established themselves wherever anything is allowed to grow. Hayle Towans are in much the same condition. Lelant Downs, across the bay, are now a golf course, where flowers can at least survive in pockets.

The fauna of the dunes has not been studied in any detail. Butterflies are plentiful, particularly the Nymphalid species (peacock, small tortoiseshell, red admiral and painted lady), the common blue and the small copper. The cinnabar moth is a common dune species whose caterpillars are often seen on ragwort. The six-spot burnet, which flies by day, and the wainscot moths are also frequent. Several grasshoppers and their relatives are present: common field grasshopper in drier places, the ubiquitous meadow grasshopper on damp slacks and the mottled grasshopper on fixed dunes; the great green bush-cricket occurs in shrubby places, and the inconspicuous little groundhopper mainly in dune slacks. Bumble-bees visit sea holly and bindweed in search of food. Spiders are abundant, a wolf spider, which makes silk-lined burrows, particularly so.

The lime content of the sand allows for a variety of Molluscs: pointed snails (*Helix aspersa* and *H. nemoralis*) and three species of banded *Helicella* snails which are prominent in most dune systems. Except for rabbits, which maintain the short turf of the fixed dunes, mammals are not numerous. Birds prefer the landward margins, though song thrushes feed throughout the dunes, where they use the occasional stone as an 'anvil' on which to break open the shells of snails. St Ives Bay and the Hayle estuary which are partly surrounded by the Hayle-Gwithian dunes, are among the best places in Cornwall for observing the wading bird, which winter in the area or rest there while passing through on migration.

Beyond St Ives Bay the north coast enters the granite land of Penwith. It continues to Land's End and then turns first south and then east at Gwennap Head, where south Cornwall may be said to start or finish depending upon which way you happen to be looking. Then comes Mount's Bay and the cliffs of the Lizard peninsula, all part of the south coast of Cornwall. But Penwith and the Lizard are such special areas that they have been given a separate chapter, so that here the south coast is taken as running from Helford River to Rame Head, Penlee Point and the Tamar estuary. This great sweep of coast, along the western extremity of the English Channel, may best be described as a series of bays, open estuaries and salt-marshes, where most of the rivers of Cornwall run down to the sea, interspersed with fine headlands, which compare well with those of the north coast.

The banks of most of the estuaries and rivers are at least partly wooded, with woods much closer to the sea than in north or north-west Cornwall; in places — parts of St Mawes inlet, for example — trees overhang the beach. Between the headlands, cliffs are less abrupt and more vegetated. There are fewer bare rock faces and only quite small areas of blown sand, indicating reduced exposure. The climate is milder. The plant-life of the cliff-tops and cliff-slopes is more lush, though many of the same species occur.

There are fewer nesting seabirds on the cliffs of the south coast. The only site which compares with the best on the north coast, Penwith or the Lizard is an off-shore rocky inlet about half-a-mile east of Nare Head — one of Cornwall's many Gull Rocks, sometimes called 'the Gray'. It supports

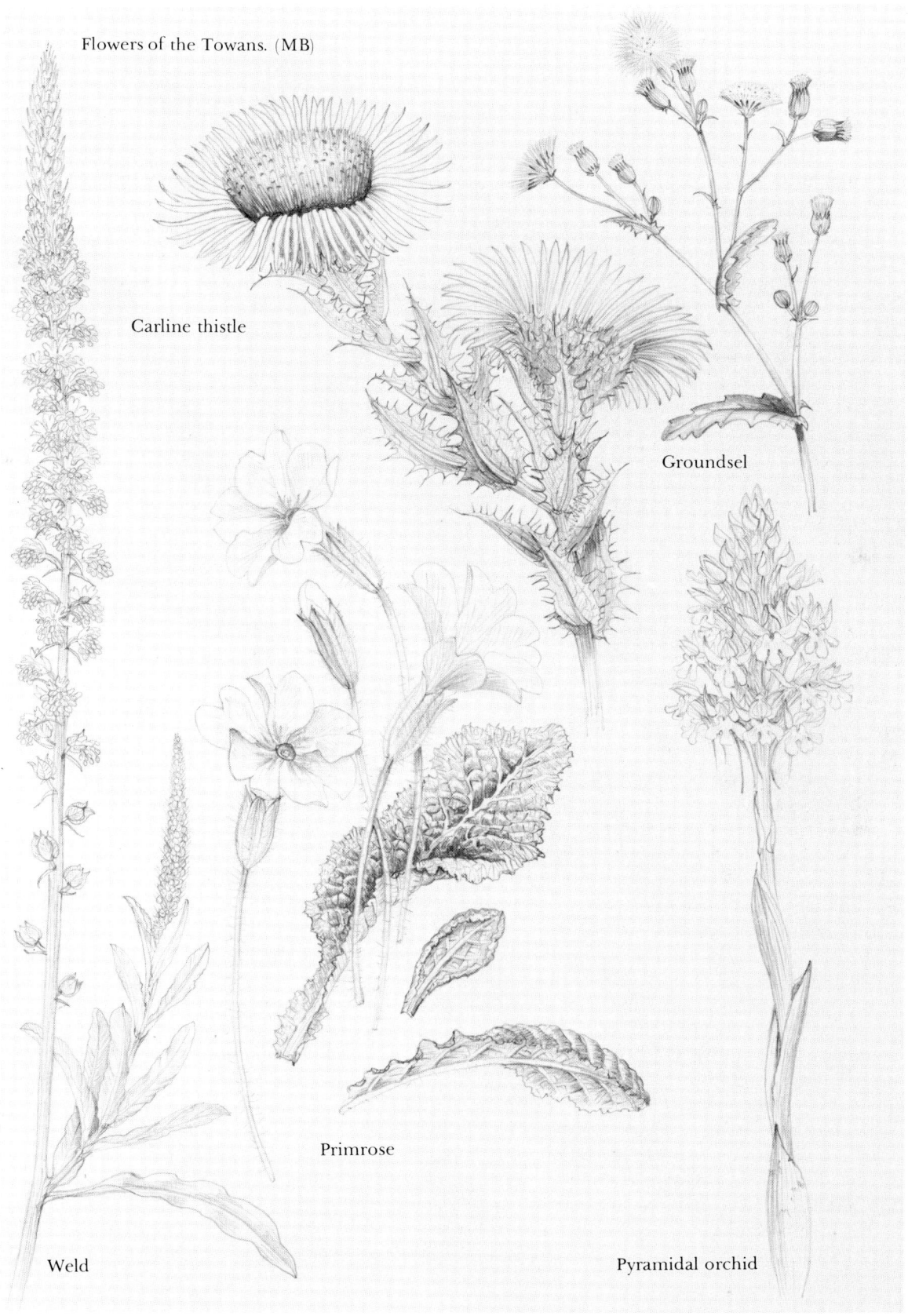

Flowers of the Towans. (MB)

Carline thistle

Groundsel

Primrose

Weld

Pyramidal orchid

the largest breeding colony of cormorants in Cornwall, a few shags, a large number of breeding kittiwakes and the only breeding auks (razorbills) east of the Lizard. Fulmars in small groups and oystercatchers in isolated pairs breed all along the coast; the former are most numerous in Veryan Bay. There are plenty of herring gulls and quite strong colonies of kittiwakes, as at Gorran Haven, but few other gulls breed. There is excellent cover and protection for many small birds, and woodland birds are plentiful close to the sea: for example, 36 species breed in Keverel wood near the mouth of the River Seaton.

In other respects there are few marked differences in the fauna. Woodland animals of all kinds are more plentiful with badgers, for example, finding ideal conditions. Grey seals do not breed, though sightings are relatively frequent. Among what are simply random observations of some of the invertebrates to be found on or near the south coast are: the keeled orthetrum dragonfly, a fairly common species, and the scarce small red damselfly on a stream near St Austell; the red-shanked carder bee, which is not common anywhere in Cornwall, on Nare Head; the grey bush-cricket, which lives in compact colonies, also on Nare Head and on the cliffs above Gorran Haven where the tawny cockroach has also been noticed; the speckled bush-cricket, which appears to lack wings and is green in colour with brown spots, in gardens and woodland verges; the rarest and largest of the British *Segestria* spiders, *S. florentina* which lives in silk tubes, has been found at Fowey; in late summer large numbers of silver-Y moths fly in from the continent.

Helford River, perhaps the most picturesque of Cornwall's estuaries, is in complete contrast to the bare cliffs of the north coast. The wooded valley of Gweek Drive connects with the head of the estuary, which is joined by numerous steep-sided creeks and edged by woods including the ancient Merthen oakwood, and others which have hardly been touched for a hundred years. Beside the water are tidal mud-flats used by waders and waterfowl. The entrance to Helford River is guarded by Nare Point to the south — rock samphire, haresfoot clover and an exceptionally fine display of orange lichen — and Rosemullion Head to the north. Rosemullion is a slate and shale promontory with low but vertical cliffs, steeply sloping scrub above and grazing land being overtaken by bracken. In spring there are great carpets of primroses and bluebells and grassland coloured by violets among them the heath dog violet; in places thick blackthorn scrub — with sycamore, elder and festoons of black bryony — stretches down to the rocky shore. From Manacle Point, which lies some way to the south of Helford River, it is possible on a clear day to see as far as Rame Head, the eastern extremity of Cornwall's south coast.

Across Falmouth Bay, Carrick Roads and the entrance to the Fal-Ruan estuary complex are the twin peninsulas of St Mawes and Roseland. Further on, beyond the headland of St Anthony-in-Roseland, are two interesting stretches of coast: the first between Greeb Point and Portscatho, the second including Gerrans Bay and Nare Head (331 feet) whose 'cliffs bristle with slatey fangs'. The former consists of about a mile of cliffs of Devonian age, facing almost due east and bearing a loose shale soil, and an extensive raised beach with many wet flushes. There is a good mixed flora, consisting mainly of familiar plants but with two colonies of the rare wild leek, one on a Cornish hedge near Portscatho, and a considerable amount of the narrow-leaved everlasting pea closer to Greeb Point, where the cliffs are lower and merge with the raised beach. Sea spleenwort is frequent in crevices in firmer parts of the cliffs. There is another raised beach along Gerrans Bay, and on Nare Head itself are exposures of greenstone and pillow lava, reminiscent of Pentire, among the slates — such outcrops of igneous rock are unusual in south Cornwall.

Beyond Nare Head and the bird islet of the Gray is Camel's Cove — the Straythe on most maps — where the cliff-top is much broken by quarrying. The geology is complex, with shaly rocks in the middle of the cove, bare greenstone on Manare Point, bevilled cliff-faces and yet more raised beaches. Where soft shales occur, cliff falls are frequent, the resulting scree supporting such plants as foxglove, kidney vetch and sea campion. Bracken has taken over in places; and there is also a large mass of an exotic New Zealand climbing shrub *Muehlenbeckia complexa* and of *Escallonia*. Little

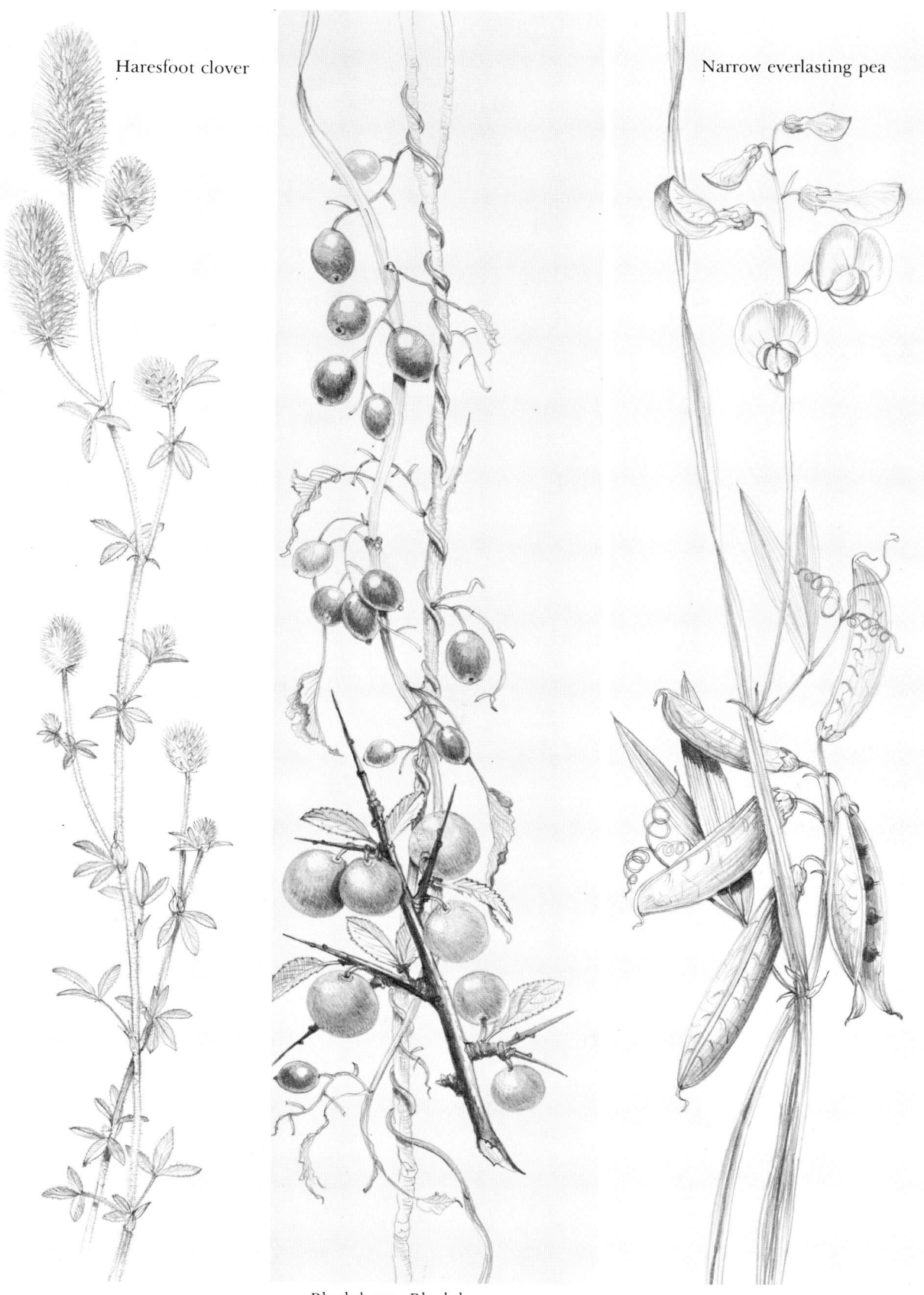

Haresfoot clover

Blackthorn, Black bryony

Narrow everlasting pea

robin and ivy broomrape used to grow on these cliffs, and the wild maidenhair fern which once occurred is thought to have been lost through cliff-falls; halbert-leaved orache, sea radish and sea beet grow nearer to the sea. At the eastern end of Veryan Bay is Greeb Point, which is volcanic ash, and the great slate promontory of Dodman Point, where there is a well preserved Iron Age fort. Exposure has produced extensive areas of impenetrable thicket, which provide food and valuable cover for small birds and insects.

The coast turns almost due north to Dodman Point. Beyond Gorran Haven, which is marred by overmuch recent development, Mevagissey, the greenstone rock of Black Head and St Austell Bay are Par sands and what remains of the marsh, still the haunt of numerous wetland birds. At the western end of the sand there used to be an old rubbish dump which was until recently a fertile hunting ground for botanists; indeed it may well have been the site of golden dock, reported by Davey in 1909 but not seen since. The site is now a caravan park. Round Gribbin Head is Menabilly, in a wooded valley near the coast, and Fowey which is cut off from Polruan by its river and estuary. Inland are the creeks of Penpoll, Pont Pill and Lerryn, all with wooded banks, predominantly of scrub oak. Lerryn drains the Trust's small woodland reserve at Pelyne about six miles inland. This includes several different habitats with deciduous woodland, a small conifer plantation, a marsh, a stream and a small disused lead mine.

Maidenhair fern. (MB)

Between Fowey and Cawsand Bay (part of Plymouth Sound) the cliffs are mostly Dartmouth slates, which often display red and green colours and which are the oldest known sedimentary rocks in Cornwall; in places they are veined with quartz. The coast between Fowey and Looe Bay is relatively uniform, with the line of cliffs broken by a few small coves and occasional high points such as the hill (447 feet) above Pencarrow Head; wild maidenhair ferns grow in some abundance on cliffs east of Polruan. There is no woodland in the immediate coastal area, the main habitats being scrub, pasture and arable. However, Polperro West Cliffs, though essentially a scrub area on sloping cliff-land, is among the best botanical sites in Cornwall, with least and lesser bird's-foot-trefoil, large cuckoo-pint, musk storksbill and the uncommon white fumitory; there are also a rare moss, *Grimmia subsquarrum* and a rare liverwort, *Riccia crozalsii*. As with the Dodman, the scrub and ivy-covered rock outcrops of the more sheltered cliff-faces provide excellent cover for birds. At Talland Bay, near West Looe, there are outcrops of beautifully coloured slates and a stretch of typical south Cornwall terrain with grassland, scrub and small wet valleys with a few willows and *Phragmites* reeds at the edge of the sea and a badger sett close to the beach. The slopes above the bay are a favourite haunt of wheatears.

Three valleys meet at or just above Looe. West Looe is still well wooded in places, particularly at Kilminorth where the Trust is working with the district council on a nature trail scheme. East Looe has suffered from too much clear felling and, as yet, no replanting. The Seaton valley is more heavily wooded, particularly around Hessenford, a few miles inland. Further east the influence of Plymouth becomes strong and there is little undisturbed coast-line. The slate cliffs are less precipitous and, where undeveloped, are much overgrown. Even so, the highest point (450 feet) on the south coast of Cornwall is on the Battern cliffs near Portwinkle, and the scrub-covered Eglarooze cliffs support the rare carrot broomrape — broomrapes are parasitic on the roots of other plants, usually on one particular species only; several grow along different parts of the south coast. 'There is nothing charming about Whitesand Bay,' says E.C. Pyatt, where the main road runs along the edge of the cliffs. But Rame Head is a marvellous situation and supports a good cliff-grassland flora. Within the compass of Plymouth Sound, there are the wooded slopes above Cawsand Bay and around Mount Edgecumbe where fallow deer sometimes leave the park. And the only Devonian limestone cliffs in Cornwall are at Cremyl and Torpoint on the western side of the Tamar estuary.

Razorbills on a rocky ledge. (FC)

OPPOSITE: Culm Cliffs from the air — Maer cliff north of Bude. (NT/H. Tempest) ABOVE: Coastal heathland — lone figure on the coastal footpath; (JB) LEFT: the Dizzard oak forest, (JB) and RIGHT: broken cliffs in the culm. (JB)

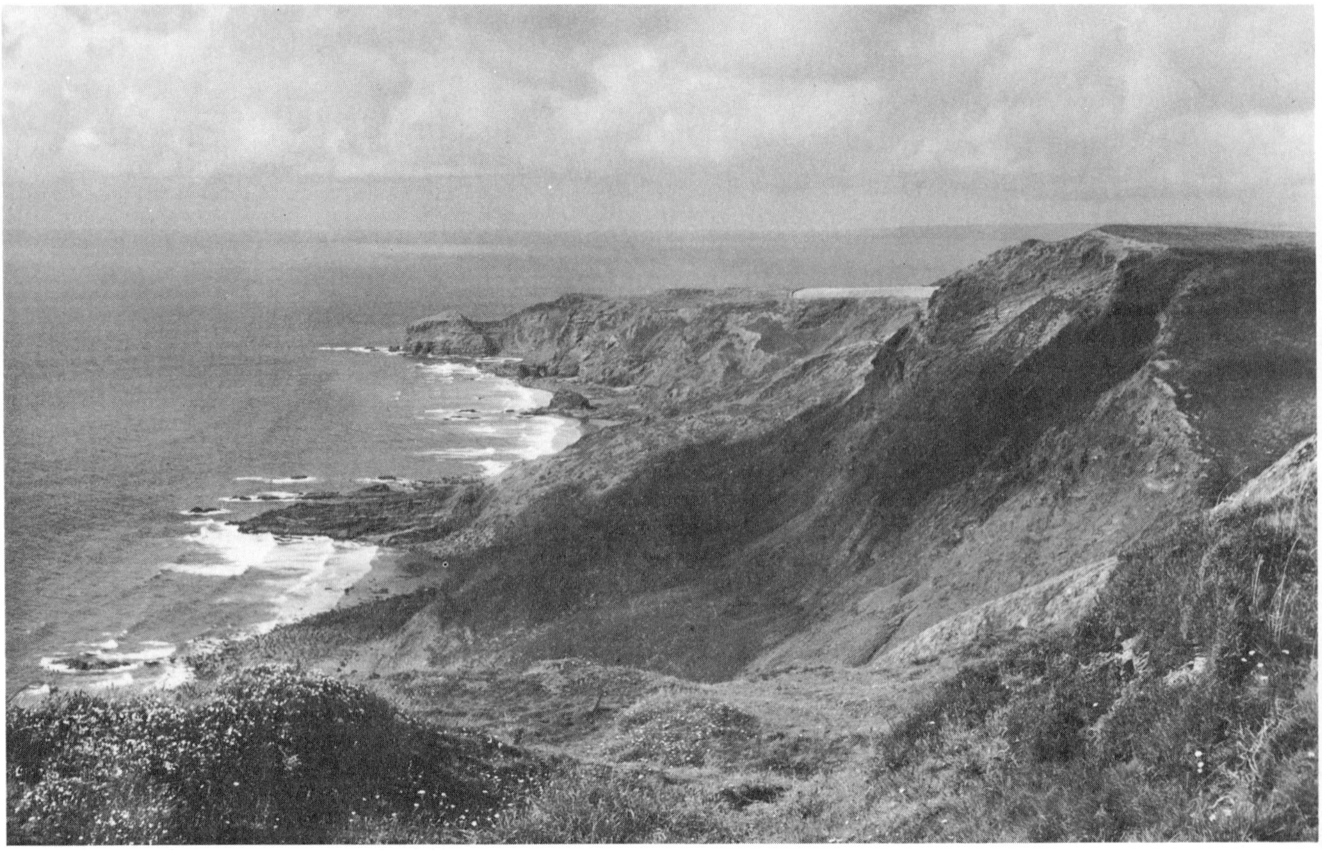

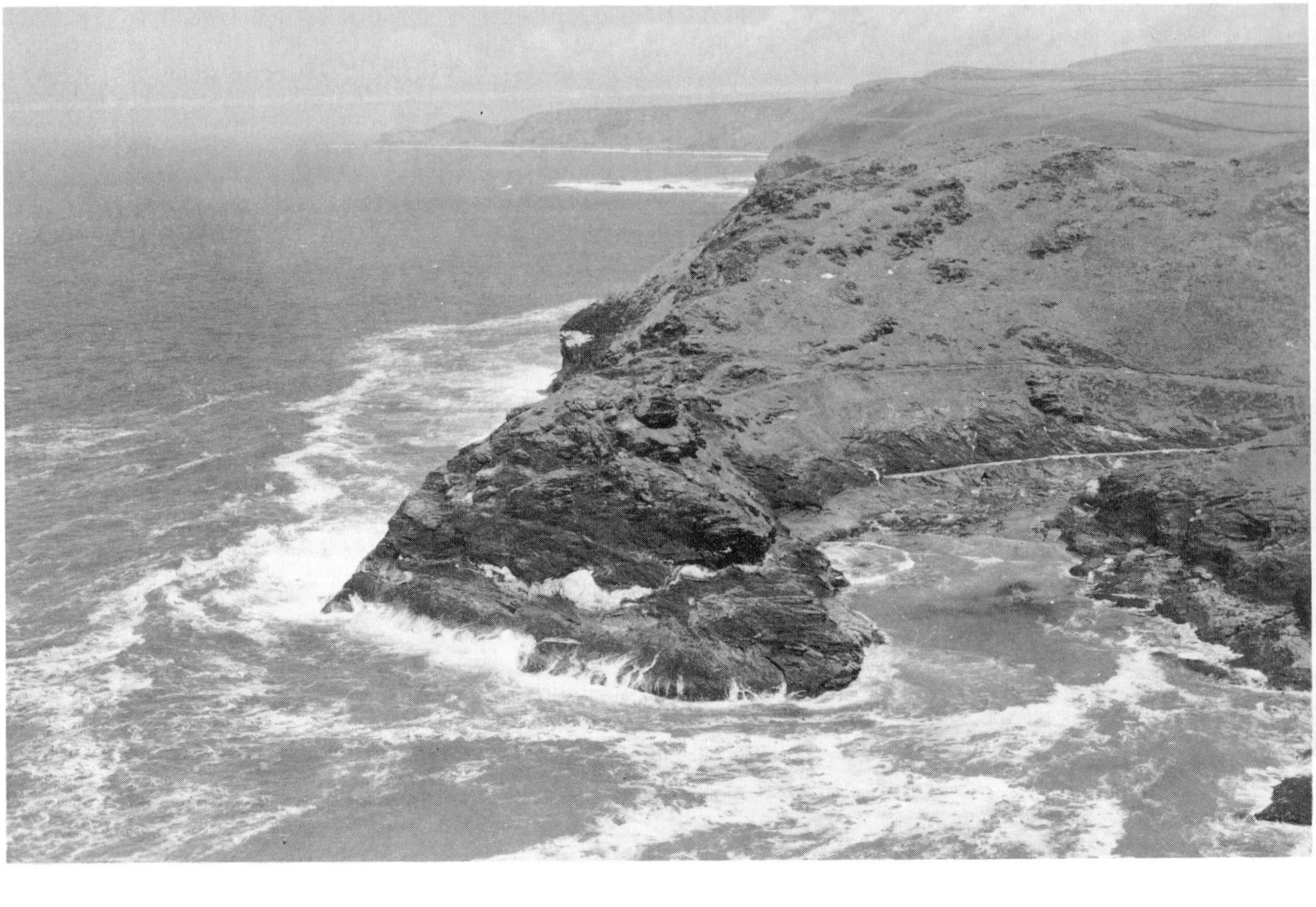

OPPOSITE ABOVE: The highest point on the coast — High Cliff with the Strangles and Cambeak; (CW/NT) BELOW: Boscastle harbour entrance — a minor break in the cliffs. (Photo Precision/NT) ABOVE: Pentire Head from the West; (CGB) CENTRE: Rumps Point Pentire — exposures of pillow lava, (CGB) and BELOW: Thrift on a north coast cliff top. (CGB)

OPPOSITE ABOVE LEFT: Rock dunes beside the Camel Estuary; (CGB) RIGHT: dune slacks; (CGB) CENTRE: development approaching the cliff top — Duporth Bay near Charlestown, (CGB) and BELOW: Porthpean Beach looking towards Black Head — a vegetated south coast cliff slope. (CGB) ABOVE LEFT: Fulmar Petrel — a bird which nests on many Cornish cliffs, (JB & SB) and RIGHT: Herring Gull — Cornwall's commonest gull; (JB & SB) BELOW LEFT: Immature Shag — Shags are more numerous than cormorants but rarely visit inland waters — (JB & SB) and RIGHT: Puffins still nest on many north coast cliffs and stacks. (JB & SB)

LEFT: Thyme broomrape — an uncommon south coast plant; (NCC) ABOVE: Stonechats are familiar birds of coastal furze, (JB & SB) and BELOW: Wheatears nest in a few places but are more often seen on migration. (JB & SB)

LEFT: The Bloody-nose Beetle is common on cliff paths, (CGB) and RIGHT: the Gorra Shield bug frequents gorse. (CGB) CENTRE: The Grey Bush Cricket, found on south coast cliffs, (CGB) and BELOW: the great Green Bush Cricket which occurs in shrubby places along the coast. (CGB)

ABOVE: Asparagus Island with Lizard Point behind, (NCC) and BELOW:
An abundance of thrift on Annet. (NCC)

Toe and Heel

Mullion Island. (FC)

The two neighbouring peninsulas, Penwith and the Lizard, differ markedly from each other and are unlike any other part of Cornwall. Both have superb coast-lines but the structure and geology are quite distinct. Away from the coast, the Lizard is a flat tableland. Penwith has a high rugged spine of moorland which runs from the edge of St Ives to Watch Croft (828 feet), and then southward to Carn Brea above Whitesand Bay 'where Cornwall ends and bathes its granite in the western sea'.

The castellated cliffs of the north coast of Penwith are well jointed so that they look like piled-up granite blocks, cut by horizontal terraces, rising one above the other. The highest points of these walls and buttresses tend to stand above the coastal plateau, a detail of cliff architecture not found elsewhere in Cornwall. The cliffs between Clodgy Point and Cape Cornwall are part of the Land's End zone of alteration ('metamorphic aureole'), created by underground movements of great masses of molten magma as they cooled and solidified millions of years ago. This is how the granite was formed as well as the hard greenstone which is present in many small exposures: Gurnard's Head and Boswednack, for example. It also accounts for the widespread mineral deposits now largely represented by numerous tips, stacks and old mine buildings.

The line of cliffs is not unbroken. Between the high points (Wicca Pillar, Zennor Head where there is a rocking stone, Gurnard's Head, Bosigran, Pendeen Watch and Bottallack) there are little sandy coves, known locally as zawns, reached by steep and narrow gullies; in places the cliffs fall sheer into the sea. The slopes and cliff-tops support plentiful, though not unusual vegetation but with mosses not frequent elsewhere in Cornwall and lichens such as the black crustaceous *Verrucaria maura* just above high-tide level and pale green *Ramalina siliquosa* slightly higher up. They also provide nesting places for numerous sea-birds: notably gulls, auks and a great many shags, but only a few cormorants. Cornwall's largest breeding colony of razorbills is on The Brisons off Cape Cornwall. There are plenty of grey seals along this coast; they frequently haul out on the Carracks, a group of off-shore rocks where they are subjected to much unnecessary human interference.

Land's End, where trampling by millions of human feet has destroyed the natural vegetation, was sold while this book was being prepared. Beyond the headland, 'a neck of land that stretches towards the west,' and southwards from it, there is a lovely part of the coast which leads past Carn Les Boel to the superb cliffs of Gwennap Head — Chair Ladders is judged by many to be the finest crag in Cornwall. Fulmars, kittiwakes and auks breed on this stretch of coast. Penwith is one of Europe's western extremities and lies in the path of the main Atlantic bird migration route, Gwennap Head (like St Ives island) being a particularly good place for observing passage migrants in flight over the sea: divers, shearwaters, skuas, petrels and gannets, sometimes in their thousands. Moreover, the sheltered Porthgwarra valley below Gwennap receives many early spring visitors (finches and warblers, for example) returning from their winter quarters; migrants also shelter there in the autumn before the long flight south. The western situation of Penwith explains the occasional presence of birds blown off course from across the Atlantic.

Between the north coast cliffs and the spine of moorland there is a narrow strip of coastal heath, and a relatively flat platform of agricultural land, cut by streams and divided into very small fields by a network of ancient granite hedges built with loose boulders. These hedges are rarely topped by shrubs, as on the south coast, but there can be a mass of colourful flowers in sheltered spots below them: red campion, foxgloves, sheepsbit, tall umbellifers and abundant ferns beside the roads. There are frequent signs of early human settlement, with stone circles, cliff castles, chamber tombs and burial grounds. Such relics are also to be found all over the moor, outstanding among them Zennor and Lanyon quoits and the Men-an-Tol — three large stones,

Ancient granite hedge, with Red Campion, Foxgloves and Umbellifers (MB)

one with a hole cut through it. Only the highest hills seem to have escaped occupation or cultivation at some time since the Bronze Age, though most of the moor has been allowed to revert periodically. The ground is rough and strewn with lichen-covered boulders; there are heathers, gorse, various ferns and bushes big enough for birds to use for shelter and nesting. A few nightjars, which are becoming increasingly scarce, still breed in Penwith. Little owls are not uncommon, and short-eared owls are regular winter visitors to the moor. Both badgers and foxes were numerous until badger numbers had to be reduced because of bovine tuberculosis. The uplands are not much grazed by sheep, but feral goats occupy some of the more remote parts.

The dark moorland soil is thick peat with numerous wet patches, supporting typical marsh plants such as bog-rush, bog asphodel and tussock sedge — insects which emerge from the pools in spring provide plenty of food for birds. Short streams flow rapidly to the north coast but the main drainage from the moor is towards the south. Water courses are longer and emerge into lush and well wooded valleys, of which Trengwainton is the outstanding example, to enter the sea by way of the coves at Porthcurno, Penberth, St Loy and Lamorna. The Trevaylor stream runs immediately east of Penzance. The Marazion stream flows into the marsh of the same name; this marsh, near which is a submerged forest, developed behind a shingle bar. It survives precariously in a much reduced state but still includes Cornwall's largest reed-bed, a favourite haunt for both nesting and wintering birds as well as a number of rare visitors such as the white-rumped and Baird's sandpipers from North America. Dunes between the marsh and Mount's Bay have been greatly damaged by human disturbance.

Southern Penwith is protected from the vicious northerly winds by the moor which slopes down gently on this side. The result is good agricultural and horticultural land and a lush flora on the coast. There are fine cliffs, and crags such as the jagged headland of Treryn Dinas, near Penberth Cove, but the thick vegetation provides good cover for small birds. This is the part of Cornwall where commercial flower-growing first developed and still continues. The Trust has an interesting little reserve, Kemyel Crease, a short distance east of Lamorna, an important centre of the trade. The reserve is an artificial wood on steeply sloping ground running down towards the rocky beach below. It is divided into minute plots, formerly used for growing flowers and early

Thrift

Wood avens

Honeysuckle

Kemyel Crease (MB)

potatoes, protected by a variety of shade trees both deciduous and evergreen. These now protect a number of ferns and flowering plants normally found in both coastal and woodland habitats: thrift and sea campion, for example, with wood avens, ribwort plantain and honeysuckle. About two miles inland is Drift reservoir, over which the Trust has a reserve agreement with the Water Authority. There is a wide range of botanical interest in this part of Penwith, from Maidenhair ferns in caves along the coast to exotics which have escaped from gardens and survive because of the damp relatively warm climate: montbretias, green alkanet and acanthus, for example. There are fewer species of butterflies and other insects in this part of Cornwall than there are further east.

Between Marazion, in a sheltered part of Mount's Bay, and the Lizard is a short stretch of coast on either side of Rinsey Head. It includes Praa sands, the Rinsey cliffs and, near Porthleven, the 'Giant's Stone', a fifty-ton erratic block of highly polished gneiss of a type found nowhere else in Britain; it is believed to have been carried to the site on floating ice. The Rinsey cliffs are the most easterly extension of the Penwith granite. Near the centre of the cliffs, above Porthcew Cove, an old mine engine-house and chimney-stack stand isolated and surrounded by heather. Abandoned for more than a century, these are now the property of the National Trust, which has safeguarded the building and capped the open shaft over which it once leant perilously.

The Lizard peninsula is usually thought of as commencing at Helston or the Helford River. Geologically and botanically, however, the limit is the boundary fault which runs from Polurrian

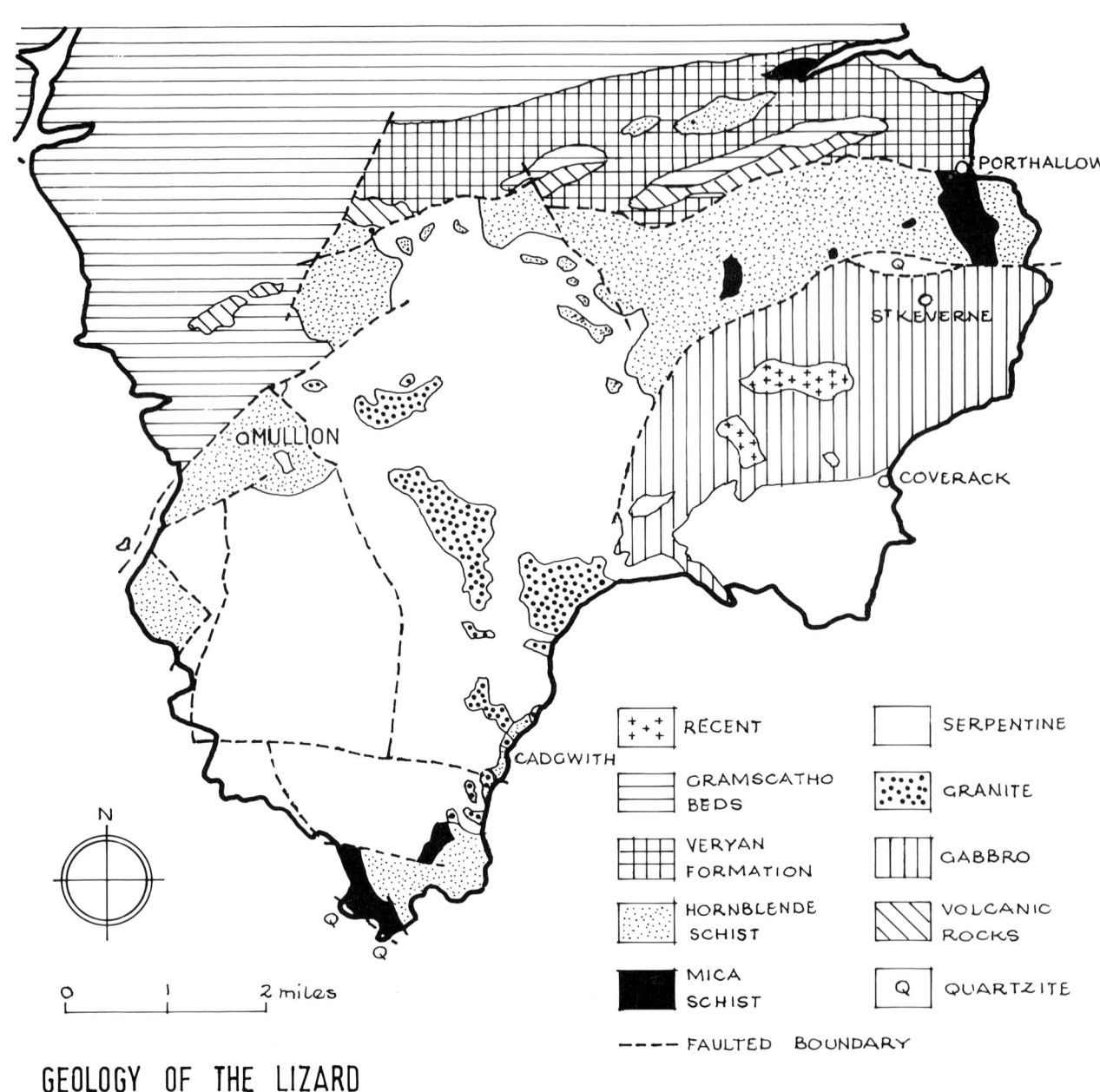

GEOLOGY OF THE LIZARD

Legend:

RECENT

GRAMSCATHO BEDS

VERYAN FORMATION

HORNBLENDE SCHIST

MICA SCHIST

SERPENTINE

GRANITE

GABBRO

VOLCANIC ROCKS

Q QUARTZITE

---- FAULTED BOUNDARY

N

0   1   2 miles

Labels on map: PORTHALLOW, St KEVERNE, COVERACK, MULLION, CADGWITH, Q, Q

Map of the Lizard. (PP)

Cove, near Mullion, to Porthallow on the east coast. North of the fault there is good Cornish farm-land on the underlying killas. South of it lie the ultra-basic serpentine rocks, hornblende and mica schists, boulder strewn gabbro with occasional outcrops of red Kennack granite and deposits of pliocene gravels. There is no escarpment. The fault is revealed by differences in the vegetation and land use.

Southward of Loe Pool, coastal rocks change first to schists, then to serpentine, and back again to the schists of Lizard Point, with an area of blown sand, badly damaged by human pressure, at Gunwalloe Church Cove. Mullion Island, a Trust reserve, is an important breeding area for sea-birds. There is Soap Cove where soapstone was once quarried, and the Rill from which the Spanish Armada was sighted in 1588. The cliffs and fragmented rocks of Kynance Cove are green with the colour of serpentine, veined red and in some places encrusted with yellow lichens. Asparagus Island, at the edge of the cove, is so named because of the prostrate wild asparagus which grows upon it. The dark cliffs of Lizard Point, below which are caverns and jagged off-shore rocks, are of mica schist decorated by cascades of exotic mesembryanthemums.

East of Lizard Point the cliffs change to hornblende schist and then to serpentine. There is the great fallen cave known as Lion's Den and, further north, the Devil's Frying Pan, 'an awesome place where the roof of a large sea cave collapsed in 1868'. There are excellent exposures of serpentine, an intrusion of red granite which extends inland to Goonhilly Downs, and the serpentine Caerverrack reef jutting seaward from Kennack sands. The sharp promontory of Carrick Luz is the site of an Iron Age fort; inland the level plateau is of clay overlying serpentine, with many prehistoric relics. The serpentine continues northward beyond Black Head to Coverack but thereafter, cliffs are mainly of gabbro until the boundary fault and the grits and shales of the killas are reached. North of the fault the cliffs are lower; not far away are the typical south Cornwall creek of Gillan Harbour and Helford River.

The Lizard plateau, which slopes gently downwards towards the south, lacks trees as well as marked landscape features. It is bisected by the road from Helston, to the west of which are the Mullion cliffs, Predannack downland — and airfield unfortunately — and Lizard Downs between Kynance Cove and Lizard Point; this whole area is of outstanding botanic interest, the result of unusual soils, mainly on serpentine rock, and the extreme south-westerly situation. A section of the National Nature Reserve and two of the Trust's reserves are in this area. Hayle Kimbro and Ruan pools, the two best serpentine-based pools, lie between the airfield and the road. They support aquatic and marsh vegetation, a number of wetland birds and are spawning grounds for amphibians: common frog, common toad and palmate newt. This is the only newt known to survive in Cornwall — it appears that both common and crested newts may

Garden escapes. (MB)

Montbretia

Green Alkanet

have occurred during the last century. These pools attract numerous insects. The invertebrate life of the Lizard is not exceptional but the scarce libellula dragonfly occurs on these pools and at only one or two other sites in Cornwall. Among spiders, the very rare *Gnaphosa occidentalis* has been found as has the scarce *Clubiona genevensis* which occurs under stones at Kynance Cove and not many other places.

The heart of the Lizard pleateau is occupied by Goonhilly Downs, a stretch of wet heath on the serpentine. There are two more large pools (Bray's Cot and Croft Pascoe), both of which have developed in old quarries, and an out-of-character conifer plantation. These downs are partially protected by the National Nature Reserve, as are the adjacent Traboe Downs, which are largely underlain by hornblende schist.

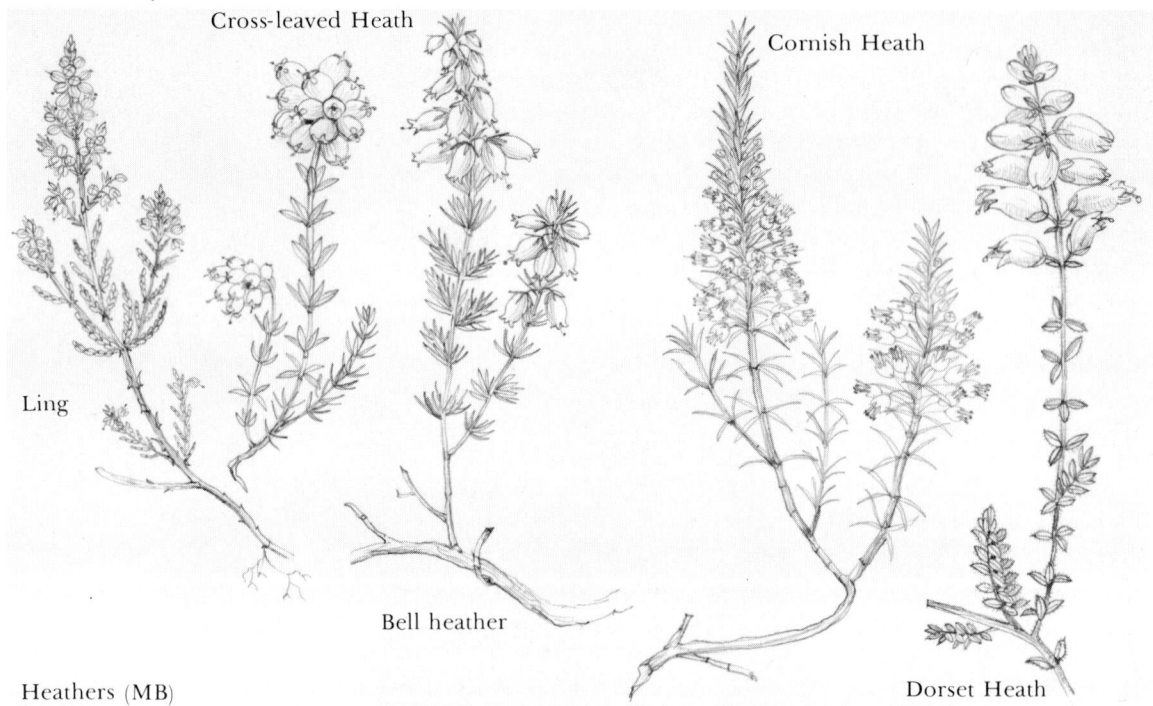

Cross-leaved Heath

Cornish Heath

Ling

Bell heather

Dorset Heath

Heathers (MB)

Further east, towards St Keverne, are Main Dale and Crousa Downs, where deposits of gravel occur. Both areas are, for the most part, boulder-covered heathland on gabbro and support a wide range of plant associations which include the rare Cornish heath, the most characteristic of all Lizard plants, whose flowers have chocolate coloured anthers and grow on leafy spikes. As well as the typical flora of the Lizard, with plants that are usually associated with calcareous soils elsewhere, the gravels support several of the more generally familiar heathers. The peculiarity of this part of the Lizard is that there are acid and alkaline soils close together, and they support plants (including mosses and liverworts) with these different soil requirements. On Main Dale there is, for the Lizard, an unusual amount of woody vegetation, with thickets of willow and blackthorn. Streams and marshes lead to a different flora again with common butterwort and the sedge, *Cladium mariscus,* abundant on the fens of East Anglia but localised in south-west England.

Although the Lizard has long been known as one of the classic botanical areas of the British Isles — its particular pride, Cornish heath, was recorded by an early botanist, John Ray, in 1667 — the various heathland habitats have recently been redescribed in much greater detail than formerly. 'Tall Heath' occurs mainly on the flatter and wetter Goonhilly Downs, where the poorly drained alkaline soils are underlain by serpentine. Cornish heath is abundant, other characteristic plants being purple moor-grass and bog rush. There is some Dorset heath, which has been greatly reduced in its native Dorset because of the wholesale destruction of heathland in that county during recent years; the conservation of Cornish sites, which are by no means numerous, is therefore of special importance. 'Short Heath' prevails on slightly higher ground, particularly the Crousa Downs gravel beds. The impression is of more stunted growth with western gorse, ling heather, bell heather and cross-leaved heath; the rare capitate rush occurs in shallow pans which tend to dry out in summer. 'Mixed Heath', with common gorse in association with Cornish heath, is also widespread on well drained sites and the sloping ground which leads down to a number of coves. There are also subsidiary heathland habitats including 'Maritime Heath' in exposed and rocky west coast situations with thrift (a Lizard variety), abundant spring squills, and ling heather as the dominant plant.

Taken as a whole, the Lizard vegetation is considered to be the best representative in Britain of a Lusitanian flora (plants found in the warm and wet climates of the west and south-west coasts of Britain, Ireland, France, Portugal and Spain) which may be exemplified by pale butterwort. Not only is there an exceptional concentration of plants rare in Britain and plants rare in the rest of Cornwall, but many common species also grow in great profusion. Most of the dominant species have been mentioned. Among others are seventeen of Britain's twenty clovers including three which are extremely rare: large Lizard, twin-flowered and upright clovers. Some rare plants have survived in shallow peaty hollows and along old cart-tracks across the heathlands; among these are pigmy rush which occurs nowhere else in the British Isles, pillwort, wild chives, pale heath violet and the three-lobed crowfoot, which is one of the rarest members of the buttercup family. Among orchids are the early marsh orchid growing on Goonhilly Downs, fragrant orchid which is often plentiful when associated with Cornish heath, and the green-winged orchid on Lizard Downs.

Sand quillwort and fringed rupturewort are to be found in some 'short heath' areas and on rocky outcrops along the west Lizard coast, with wild asparagus and other prostrate plants including water figwort, lesser meadow rue and the prickly sedge. As in so much of Cornwall, a wealth of plants grows on old walls and hedges including white stonecrop and Bithynian vetch which, though local, are not confined to the county.

Most habitats, including heathland, benefit from a reasonable pressure of grazing provided it is not accompanied by too much 'pasture improvement'. Lizard soils are not suited to intensive agriculture — soils on serpentine are deficient in calcium, potash and phosphate but contain toxic nickel and chromium — though this has not prevented some unfortunate reclamation during the last thirty or forty years. Other developments have also occurred, including the earth-satellite station on Goonhilly Downs and the Predannack airfield; the result is that the

Fragrant orchid

Green-winged orchid   (MB)

heathlands are much fragmented already. The main immediate threat seems to come from the possibility of afforestation, which could change the whole character of the peninsula and destroy its botanical importance. There already are plantations north of the boundary fault and two to the south of it — on Goonhilly and Crousa Downs. Any extension of these would be most unfortunate. Small scale quarrying has been going on for generations and does little harm, so long as it remains small.

As yet, tourism makes little impact on the heathlands, though careless picknickers sometimes start fires, but the Lizard is vulnerable to any marked extension of the holiday trade, which can be very destructive to fragile habitats. In spite of the National Nature Reserve, the Trust reserves and important holdings of the National Trust (covering some 2,000 acres in all) there is need for intelligent overall conservation management. While many rare species remain — some of which must be classed as threatened — sea knotgrass, pennyroyal and slender bird's-foot-trefoil are known to have become extinct since 1945; it is feared, moreover, that fir clubmoss, yellow-horned poppy and one or two other species may have gone the same way.

Except for a few woodland species, most of the breeding birds of Cornwall nest somewhere on the Lizard or have done so in the recent past. Montagu's harrier used to nest regularly on Goonhilly Downs, but has not done so since the 1930s. The old peregrine eyries on the coast are no longer used, but this fine raptor is gaining ground elsewhere in Cornwall and might well return. Red kites are believed to have bred early in the last century, and a pair was seen at Mullion as recently as 1952. While there is no reason to anticipate the return of this rare bird, the Lizard, with numerous small mammals and amphibians, is undoubtedly attractive to birds of prey. Nightjars still nest on the peninsula, as does the wheatear which seems to be another decreasing species. But the main ornithological importance of the Lizard is as a nesting area for sea-birds. The best sites are on the west coast, notably Mullion Island and the Vro where large numbers of shags, herring gulls, greater black-backed gulls and kittiwakes breed regularly with a few cormorants, guillemots and razorbills. Other worthwhile centres are between Kynance cliffs and Polpeor Cove, where fulmars nest, and Black Head on the east coast where there is a large colony of herring gulls.

Shags nesting at Hell's Mouth, Lizard — looking straight down from the clifftop. (FC)

ABOVE: Porthmeor Cove and the north coast of Penwith — moorland and coastal strip showing the pattern of small fields. (H.Tempest/NT) BELOW: Bosigran — granite cliffs and a Penwith zawn. (NT)

ABOVE: Marazion Marsh — the most extensive reed-bed in Cornwall; (JB & SB) BELOW: Kynance cliffs and flat plateau surface of the Lizard from the air. (RNAS Culdrose/NT)

ABOVE LEFT: Lizard heathland above Kynance Cove, (NCC) and
RIGHT: West Lizard Downs — the site of many interesting plants; (NCC)
CENTRE LEFT: Mullion Cliffs and Island, (NCC) and RIGHT: the rocks of
Men-te-heur near Mullion; (NCC) BELOW LEFT: An old cart track on the
Lizard Downs — site of the rare Pigmy Rush, (NCC) and RIGHT:
cultivation encroaching upon the Lizard heathland. (NCC)

ABOVE: A pool on the Goonhilly Downs; (NCC) LEFT: the rare Spotted catsear still present on the Lizard and one other Cornish site. (NCC) CENTRE: Pigmy Rush — found only on the Lizard; (NCC) BELOW: a Grey Seal 'bottling' near the Carracks.(JB & SB) ABOVE LEFT: Short-eared Owl which frequents the Penwith Moors in winter, (Donald Smith) and RIGHT: Little Owl — often seen by day in Penwith; (A. Winspear Condall) BELOW LEFT: Cocoon and caterpillar of Five-spot Burnet moth which associates with Birdsfoot Trefoil, (CGB) and RIGHT: Adult Five-spot Burnet Moth. (CGB)

ABOVE: A rocky shore — Porthgwara Cove, (CW) and BELOW: relics of early human occupation of Penwith moors — Men-an-Tol. (CW)

Wetlands:
Estuaries and Inland Waters

Loe Pool from the air with Wigeon approaching. (FC)

The climate and topography of Cornwall assure plenty of fresh water, with rivers rising on the moors and ending in tidal estuaries, streams in almost every valley and much marshy ground. The only natural lakes of any size, however, are Dozmary Pool on Bodmin Moor and Loe Pool below Helston. To compensate for this, past land-use in the form of quarrying and mining has left parts of the countryside pock-marked by artificial water-filled holes and hollows. There are also twelve reservoirs (counting the three small Crowan reservoirs as one, and including Colliford on Bodmin Moor) covering a total of 1,781 acres, and thus the major component of the still water habitat of Cornwall.

To many naturalists the particular interest of open water lies in the wintering wildfowl it attracts. In general, a straight shoreline is less attractive than a series of small bays; the more varied the verges and immediate surroundings, preferably with bushes and trees not too far away, the more interesting the fauna is likely to be. Mud around the water's edge and extensive shallows, where reeds and sedges are plentiful, attract waders, particularly those on migration, and are also used for nesting by coots and little grebes. Dabbling ducks favour shelving verges; teal and wigeon, for instance, relish plants such as amphibious bistort and water starwort. Underwater plants encourage invertebrates and attract diving ducks — tufted duck and pochard most commonly in Cornwall. Ducks rarely feed at depths greater than twenty feet so that deep water lacks wildlife interest. Fortunately human recreational use of reservoirs is greatest during the summer, when the flocks of wildfowl are elsewhere.

The first reservoir to be constructed (1819) was Tamar Lake on the Devon border, at that time part of the Bude Canal project. It was declared a bird sanctuary in 1949 and two years later became one of the eleven Regional Wildfowl Refuges in England and Wales. Though relatively small (51 acres), verges are unusually varied and the lake shallow, so that it is attractive to wildlife. The only other reservoirs constructed before the late 1950s were College (38 acres) Argal (62 acres), Bussow (7 acres) and two of the small Crowan reservoirs (9 acres). Then came the era of modern reservoir construction with Porth (38 acres) near Newquay, Drift (64 acres), Sibleyback (140 acres), New Tamar Lake (81 acres and immediately above the old lake) and finally Colliford (905 acres) above St Neot, under construction in 1981. The long, shallow, northern arm of this reservoir is destined to become a nature reserve.

While open water always attracts wildfowl, small reservoirs are visited only briefly, though a remarkably varied collection of birds has found its way to Crowan over the years. Bussow lacks surrounding vegetation, and relatively few water-birds seem to winter on either College or Argal. Both Sibleyback and Porth are regularly used for sailing, yet are visited by good numbers of

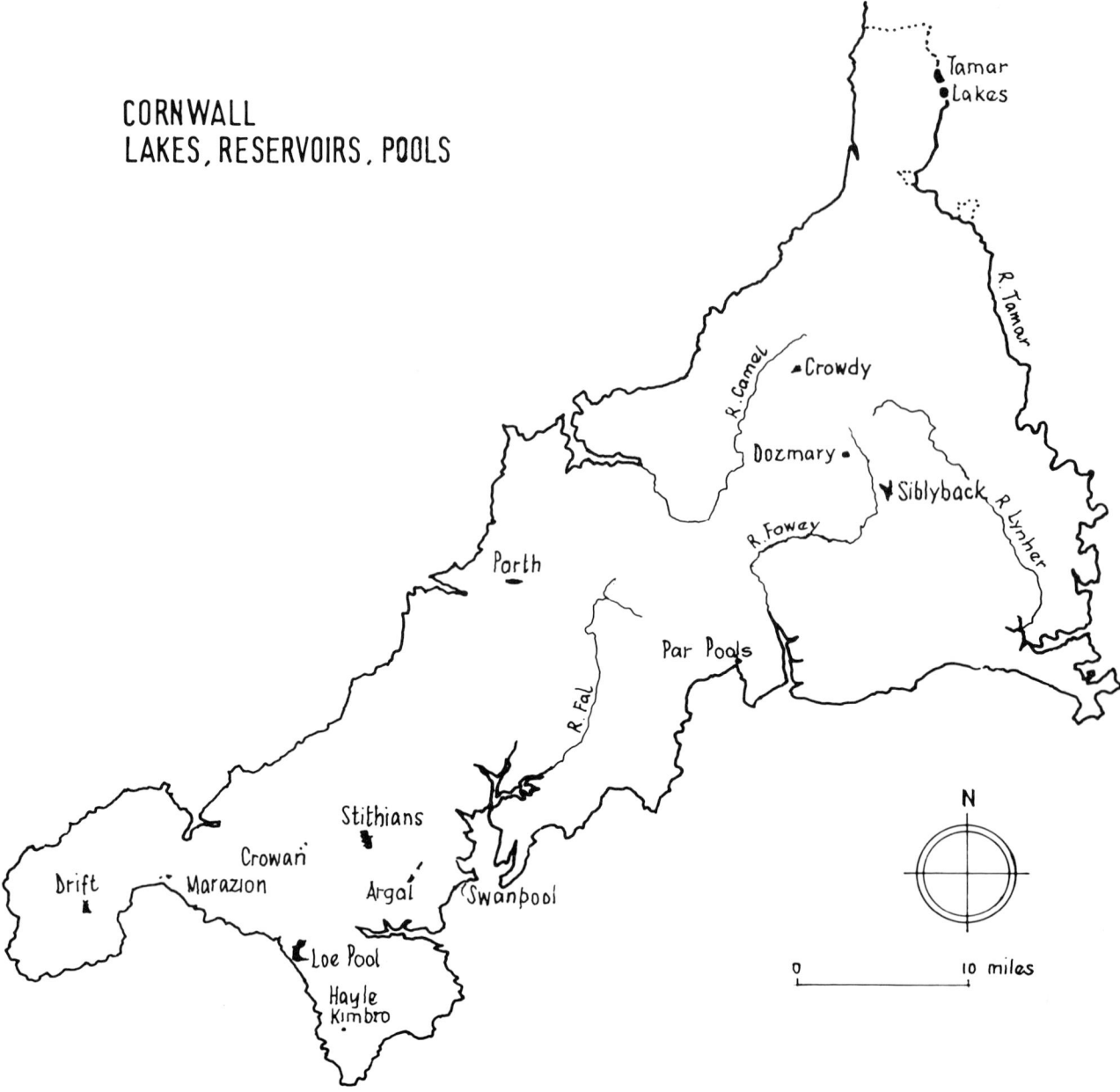

CORNWALL
LAKES, RESERVOIRS, POOLS

Tamar Lakes

R. Tamar

R. Camel

Crowdy

Dozmary

Siblyback

R. Lynher

R. Fowey

Porth

R. Fal

Par Pools

N

0   10 miles

Stithians

Crowan

Drift

Marazion

Argal

Swanpool

Loe Pool

Hayle Kimbro

Reservoirs in Cornwall. (PP)

wintering wildfowl; woods near the latter provide nesting places for many other birds including the immaculate little dipper which breeds by a stream. The Trust has a reserve agreement over Drift, the most westerly reservoir in England. There is a considerable expanse of shallow water and plenty of mud in the autumn, which makes it a useful landfall for migrating waders. Proximity to the sea explains the frequent appearance of birds which seldom leave the oceans — including at times such rarities as glaucous and Iceland gulls. Crowdy, high up on Bodmin Moor and adjoined by a typical upland marsh, attracts not only birds but numerous invertebrates as well. In many ways, however, the most noteworthy reservoir in Cornwall is Stithians, which is visited by large flocks of wildfowl — up to 400 wigeon and 200 teal at times. Relatively shallow and surrounded seasonally by wide expanses of mud, it is the county's most important inland area for migrant waders such as sandpipers, godwits and redshank.

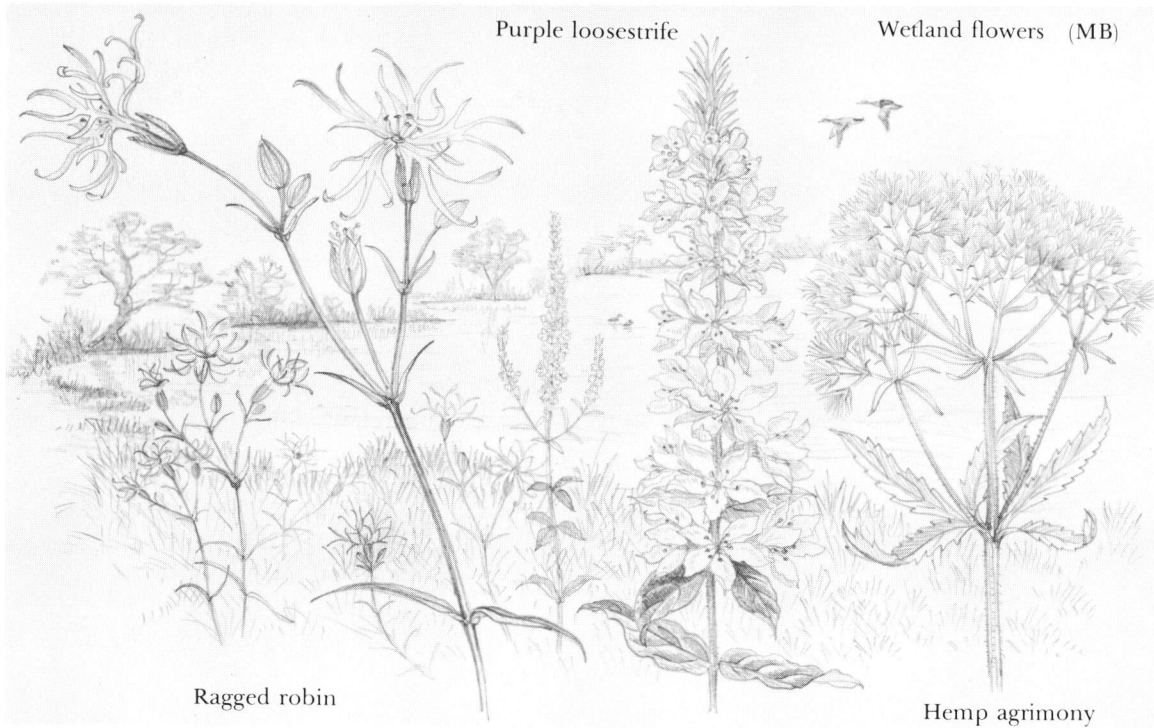

Purple loosestrife   Wetland flowers   (MB)

Ragged robin

Hemp agrimony

Because of the unusual variety of habitats, the two Tamar Lakes, which constitute a single ecological entity, provide an excellent example of the inland waters of Cornwall. The shallow arms of the old lake are fringed with trees, where many birds nest and roost, and there are beds of rushes, reeds and sedges. Colourful wetland flowers — such as ragged robin, purple loosestrife and hemp agrimony — attract bumble bees and butterflies. Aquatic vegetation includes water starwort, bog and curly pondweed, hornwort and alternate water milfoil which grows submerged with spikes of yellowish flowers projecting above the surface. The surrounds of the new reservoir are mainly grassland and meadow, where duck and coot feed, but its northern marshy extremity is well used by snipe and other migrant and wintering waders. In winter mallard, teal, wigeon, tufted duck, gadwall, pochard and large flocks of coot are always present on both lakes; usually shoveler and goldeneye are there too. Other wildfowl appear occasionally and almost anything may turn up on migration. Herons are commonplace. Bittern, night heron and all five British grebes have been seen. Tamar Lake is far from being simply a wildfowl refuge.

Cornwall's two natural lakes could not be more different. Loe Pool (150 acres) is at sea level and was formed by development of a shingle bar (composed of flint, chert pebbles and coarse sand) across the mouth of the River Cober. It winds back from the sea like the river which in fact it is, and is bordered by trees, shelving meadows and marshes. White water-lilies grow in the lake; there is a good range of wetland plants around the verges, particularly at the edge of Carminowe Creek, and strand vegetation including sea-campion and sea-purslane on the bar through which water seeps into the sea. The life of the sea shore and of the woods begins at the edge of the pool.

Dozmary Pool (150 acres including the adjacent marsh) is by contrast a lonely brooding place among high moorland hills. Commonly described as 'bottomless' and the source of several strange legends, it is in fact a shallow depression at the only point on Bodmin Moor where conditions favoured formation of a lake. One rare plant grows in the marsh: small quillwort. Seventeen species of wildfowl have been recorded in recent years, mallard and pochard being the most numerous. As well as other duck found on most of Cornwall's inland waters, greylag goose, goosander and Bewick's swan have also been observed.

Crowdy reservoir, at a slightly higher altitude (918 feet) and not established until 1972, attracts more species and more birds. Mallards, wigeon and teal are frequently seen in flocks of over one hundred; white-front and barnacle geese, red-breasted merganser and shelduck have appeared on Crowdy but not Dozmary. Whooper swans have been recorded as well as Bewick's. Several waders, including curlew, frequent Crowdy as do herons, gulls, grebes and sometimes terns. The grassy and scrubby surrounds bring other birds to the neighbourhood, and a hen harrier sometimes hunts there. It is an exceptionally attractive spot, overlooked by the wild and rugged slopes of Rough Tor. A dark wall of trees protects one corner, and huge flocks of lapwing and golden plover often probe for food on the open moor above, where the ground is broken for them by grazing cattle and horses. As well as the flowers in Crowdy marsh there are patches of gorse and grassland plants beside the reservoir, where they encourage insects and provide both food and shelter for numerous small birds. That unusual and uncommon fern, moonwort, was found there soon after the reservoir was finished, but quickly vanished under uncontrolled grazing by too many sheep.

The small lake in Pendarves reserve was dug out in the late 18th century to add to the amenities of Pendarves House. Fed by a woodland stream, it is partly surrounded by trees, with marsh and marginal vegetation beside the water. There are clumps of willows and alders with bulrushes and a riot of plants around them. Parts of the pool are becoming overgrown with water lilies, pondweeds rooted in the silt, water starwort and water crowfoot: among non-flowering deep-water plants, stonewort and algae are common. While some of this has to be cleared away, it greatly benefits the fauna of the pool which includes trout, stickle-backs, palmate newts, toads, pondskaters, whirligig beetles, river sponges, flatworms and leeches. Mayflies of several species hover above the surface, and among several dragonflies are the southern aeshna, a hawker which is rare in Cornwall, though sometimes seen flying around woodland glades. Another of the less common species recorded here is the small red damselfly.

Red Moor reserve (50 acres) below Helman Tor, was bought with help from the World Wildlife Fund as a memorial to Bill Almond, Derek Johnson and Enid Campbell, all of whom worked for the Trust with great dedication during its early years. It is not strictly a wetland area but, among several distinct habitats, there are a lake some fifteen feet deep and several ponds (formerly gravel pits and mining excavations) silted up and overgrown in varying degree. The ponds drain into each other and finally into the lake, which feeds a stream leading to the River Fowey. There is a dense willow carr, a profuse growth of lichens, some patches of *Sphagnum* bog, an area of purple moor grass and of cross-leaved heath. Such variety means an unusually interesting flora which includes cotton grass, royal fern, bog asphodel, guelder rose and three species of orchid. Red Moor is particularly rich in aquatic bugs, beetles, hover-flies, dragonflies and spiders. Among the latter there is the water spider, building silken 'diving bells' in shallow weedy ponds and living almost side by side with dry country wolf spiders. The fauna also includes palmate newts, frogs, toads, viviparous lizards, slow worms, grass snakes and adders. Several mammals, including badgers, find a home on Red Moor as do an extremely varied collection of birds from moorhen to warblers, pheasant and buzzard.

The major marshlands of Cornwall are mostly on the higher moors, but the wetter parts of numerous lowland heaths are usually boggy. An example is Retire Common, where a vein of quartz runs through the underlying killas and there is an interesting zoning in the vegetation. Bell heather grows profusely in the better drained sections while the wetter areas support sedges, bog rush, bog myrtle, deer-grass which is usually restricted to the higher moors, lesser butterfly and marsh orchids.

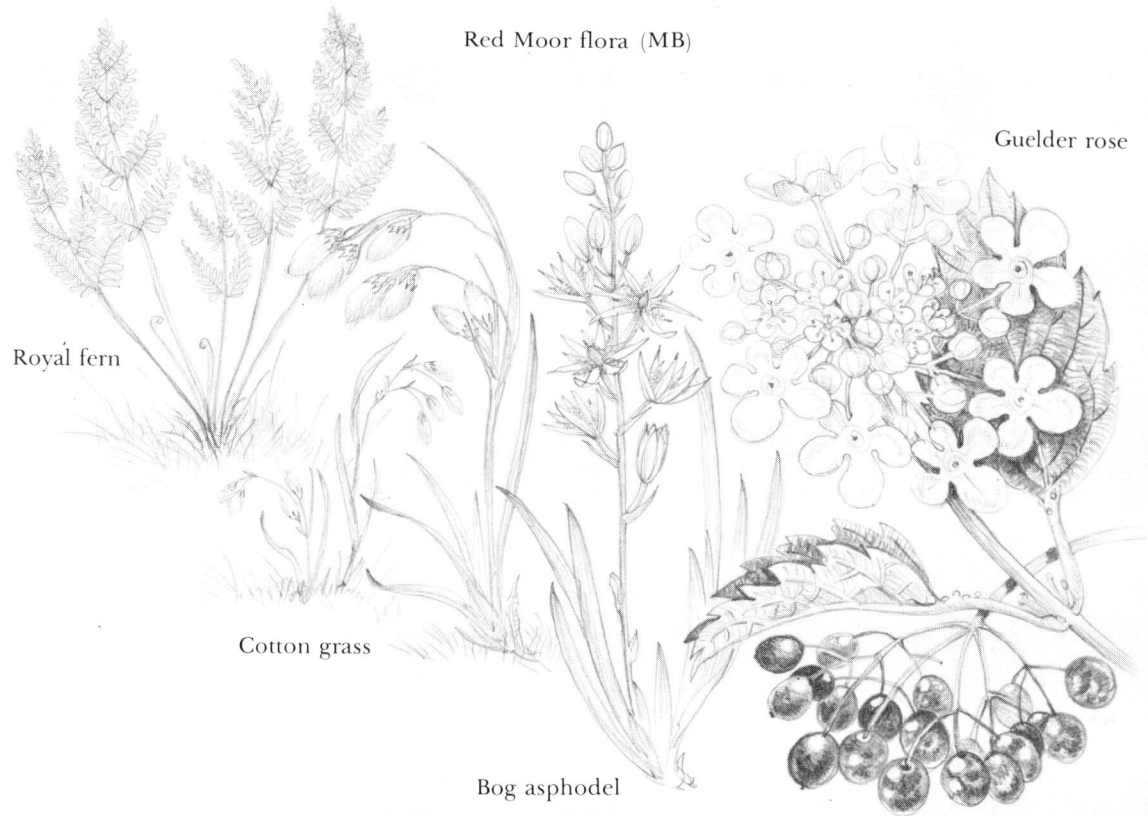

Red Moor flora (MB)

Guelder rose

Royal fern

Cotton grass

Bog asphodel

One of the best low altitude bogs in Cornwall is Ventongimps Moor, a reserve of which the Trust owns the freehold. Since the early 1960s it has gradually become drier and less acid and may have suffered pollution from adjacent farmland.

Gorse has been taking over from Dorset heath which was widespread. *Sphagnum* moss has become less frequent, as have other bog plants such as pale butterwort. The rare insect-trapping greater sundew, which had been quite plentiful, was already becoming scarce in the 1960s and has not been seen since 1975. Wavy St John's wort, another plant for which Ventongimps was noted, is now much reduced, and only a few lesser butterfly orchids remain. This marked reduction in number and variety of flora has reduced the survival potential of the marsh fritillary butterfly in this area. Management of the reserve has sought to reverse these trends by recreating the earlier, wet conditions.

Most of Cornwall's rivers flow southwards from the granite moors. The River Tamar, however, rises in marshy ground north of Kilkhampton. The Camel's main source is between Bodmin Moor and the coast, the river eventually turning sharply north to reach the sea. Hayle river takes a rather similar course much further west, passing through the Trust's Tremelling reserve (pools, marsh and woodland) on its way to the estuary and St Ives Bay. The Fal rises from the Hensbarrow granite above St Austell and was, until recently, much polluted by china clay waste as was St Austell River. The only other river to have suffered much from pollution is Red River which, for hundreds of years, has carried tin waste from the Camborne mines to St Ives Bay. Numerous streams run their short courses and find their way to the sea in the mouths, coves, bays, zawns and minor estuaries with which the coast abounds.

Ventongimps flora (MB)

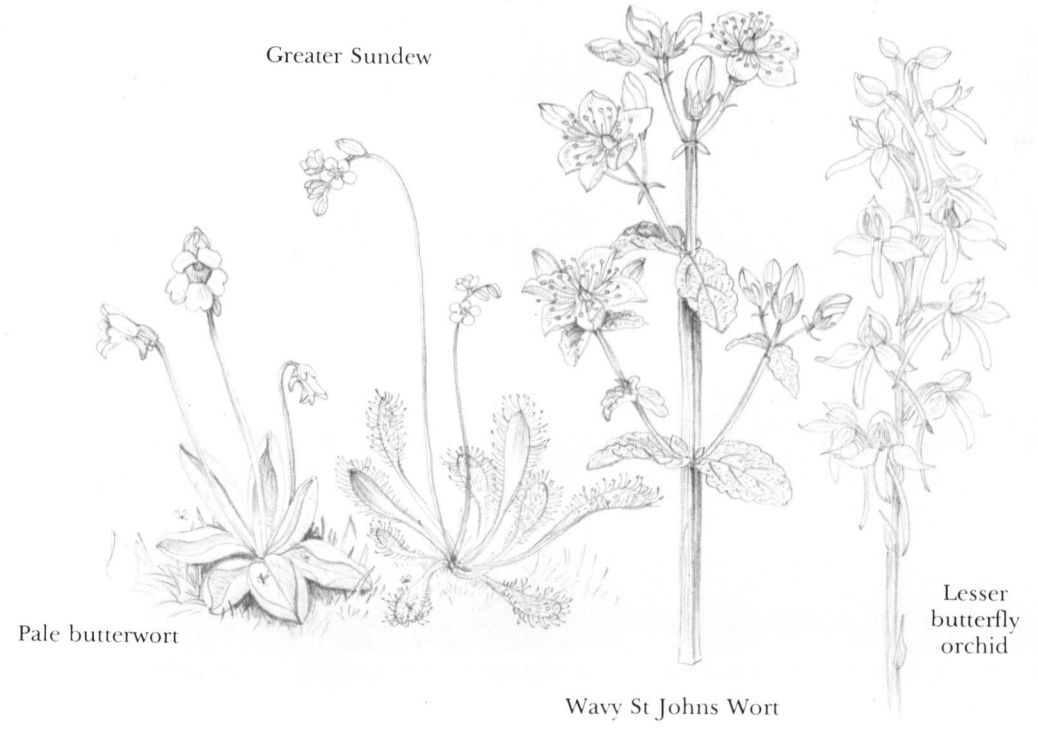

Greater Sundew

Pale butterwort

Lesser butterfly orchid

Wavy St Johns Wort

Canals have never been of much importance in the life of Cornwall. What remains of the Liskeard-Looe canal, much of which has been filled in, is now overgrown, while the canal between Par and St Blazey has been so polluted with mica-clay that few plants grow in it. The old Bude Canal is still waterway for about two miles and shares a fine open alluvial valley — where the yellow flag makes an unusually fine display in spring — with the River Neet. Trees of varying height, a marsh between canal and river which has been declared a nature reserve by the district council, reed beds and the two different waterways have made this an excellent area for birds and riverain wildlife generally. Snipe and water rail habitually winter in the reserve.

Because most of Cornwall's rivers rise in moorland bogs, the water tends to be acid and somewhat deficient in nutrients. There is a sufficient freshwater fauna even so, and small brown trout are numerous — but not the large specimens found in chalk streams. Salmon and sea trout find their way to gravel spawning beds in the Tamar, Camel, Inny, Fowey and Lynher Rivers; they seldom appear to enter the Fal. Except in relation to the extractive industries, industrial pollution has not been serious, but farm effluence has caused occasional problems. Pollution damages organisms living in the water and may destroy the vegetation upon which animal life depends. There are, however, natural indicators. The presence of kingfishers shows a sufficiency of small fish, such as bullheads and minnows; herons, wagtails and water voles indicate water clean enough to support them. Pools which have developed from past industrial activity vary greatly.

Some are almost completely sterile. Others support a flora and fauna comparable to that of Pendarves lake — rudd breed in Relubbus pools near Penzance, for instance. Excessive zeal in river management can be damaging to the wildlife. A meandering stream with sheltered bays is preferred to one which has been straightened out and converted into a glorified drainage channel. Rivers and river banks do not attract many birds or other animals unless there are trees and shrubs, a reasonably varied flora and occasional marshy verges.

The Otter Report, published in 1981, emphasises the importance of river management in relation to survival of the otter, surely one of the most attractive wild animals of our land. The report stresses the part played by river-bank trees, particularly oak and ash and sycamore, in the life of the otter, because cavities develop between their roots and provide lying up and breeding places — willow and alders are less important. The survey found positive signs of otters in the Tamar, Lynher, Camel, Fal and Neet River systems as well as in a few streams and two reservoirs. Thoroughly suitable habitat occurs in several other rivers including the Cober, Helford and Seaton. The coast, which at one time was much used by otters, was not covered adequately owing to difficulties of access. The report concludes that the area of the South West Water Authority is 'the most important for otters in southern England because most of the rivers have unpolluted waters and good bankside cover giving otters their essential food and shelter'. But pressures are increasing and otters have already disappeared from several rivers in the county. Alien mink, moreover, have now spread throughout Cornwall, and no one yet knows how successfully these two species can co-exist.

Other mammals associated with water are the rather beautiful water shrew, with its velvety black back, and the water vole. Both are widely but sparsely distributed in Cornwall. Dragonflies are the invertebrates most likely to catch the eye; among the commoner species are the golden-ringed dragonfly, a large black and yellow banded hawker, the four-spotted libellula, a darter which prefers still or slow-moving water and the large red damselfly. When the water is not too acid, wandering snails are common, and ramshorn snails are often to be found, as are small freshwater limpets on stones among the weeds; large freshwater mussels occur in a few waterways and in Old Tamar Lake. The abundance or otherwise of animal life in streams depends largely on the rate of flow. Fast-flowing streams support little plant life and therefore attract few animals except for some stonefly nymphs and caddis larvae. More gently flowing water allows plants to take root, so the fauna is more abundant.

Estuaries where rivers meet the sea are drowned valleys resulting from a gradually rising sea-level. Silt washed down by rivers builds up in estuaries and is covered twice daily by tides

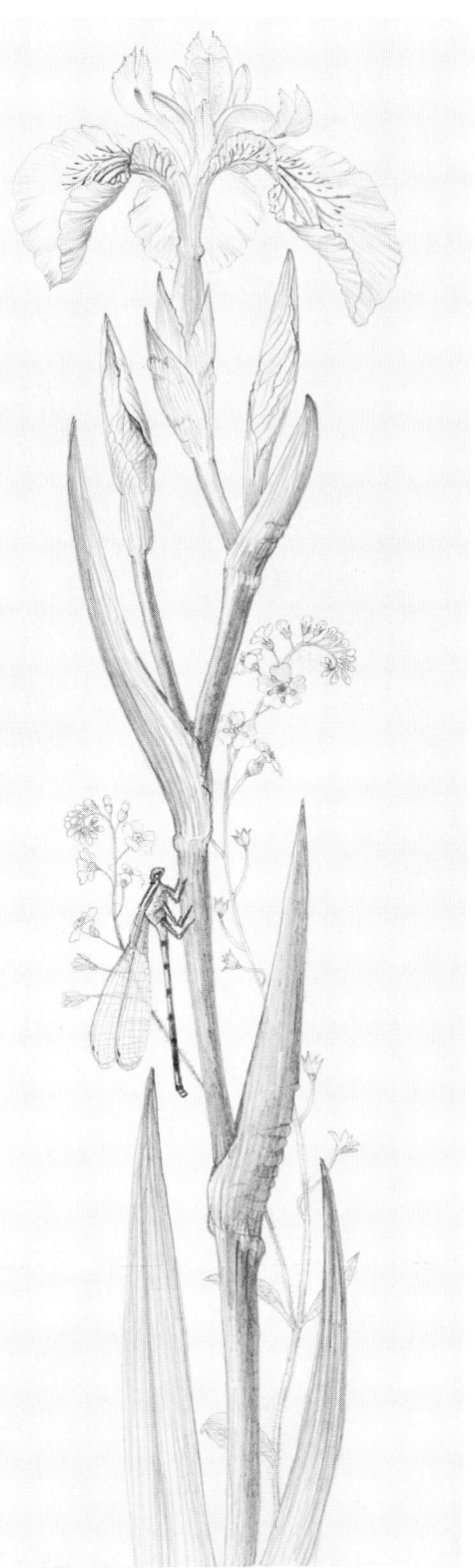

Yellow flag iris. (MB)

which bring various marine organisms with them. The result is salt-marsh, a biologically rich habitat, but one which can support only those plants that tolerate highly saline conditions. Among these are glasswort with miniature flowers and succulent stems, cord-grass which is abundant in the Tamar and Lynher estuaries, the silvery-leaved sea-purslane and the salt-marsh forms of two very familiar and attractive plants: sea-aster and sea-lavender. Numerous animals live in the mud and are present in great abundance: peppery furrow-shells, cockles, sand-hoppers, lugworms which make casts and ragworms which don't, are some examples. This is what brings birds to the mud-flats and makes them such precious places for many naturalists.

Otters on a river bank. (MB)

The only major estuary on the north coast is the Camel, an immensely varied system largely dominated by sand. The Hayle estuary is less extensive but provides for an exceptional concentration of wintering wading birds, its south-westerly situation attracting both migrants and birds seeking shelter in severe weather — it is one of the few places in Cornwall where stone-curlews have ever been reported. Hayle is an area of high tourist pressure and is constantly under threat. There are also the Gannel estuary, where birds appear in small numbers but good variety in spite of constant human disturbance, and a small estuary at Bude, where there are good growths of long-leaved scurvy-grass and sea-aster.

While there are several estuaries and tidal creeks with steeply wooded sides along the south coast, the most important areas are the Fal-Ruan-Tressillian complex and the Tamar-Lynher estuary with St John's Lake. Parts of both are Trust reserves: mud-flats beside the Tamar leased from the Duchy of Cornwall, and Ardevora and Trelonk salt-marshes on the Fal from the National Trust. The unique interest of the latter area is that it is underlain by pure white clay, brought down the river from china clay workings between Grampound and St Austell. First colonised by sea pea and sea club-rush, a normal salt-marsh flora has developed with sea-asters and sea-spurrey. The area attracts waders and numerous wildfowl, including shelduck which nests on the banks of the estuary sometimes in rabbit burrows; frogs and toads are also common. Above Ruan-Lanihorne, tidal forest vegetation, a most unusual habitat, has evolved by natural succession from salt-marsh. The woodland is mainly willow, alder and oak with sessile oak (stalkless acorns) once coppiced to produce bark for the tanning industry.

The Tamar estuary on the Devon border has the largest area of mud-flats in the South-west. It stretches up the river for twelve miles above Plymouth and includes, on the Cornish side, St John's

Lake south of Torpoint, the Lynher or St German's estuary and Kingsmill Lake. The whole area is visited by large numbers of waders including curlew, dunlin, knot, redshank, greenshank and very large flocks of golden plover. For the striking avocet, with its long upcurved black bill, the Tamar above Cargreen is the most important English wintering place. The Lynher, which enters the Tamar a short distance above the bridge, supports several different estuarine habitats ranging through the full sequence from fresh to salt water. Sheviock wood immediately alongside the estuary accommodates 38 species of breeding bird including heron, buzzard, tawny owl, nuthatch and tree-creeper. Three species of *Zostera,* eel-grass or grass-wrack, grow in the Tamar estuary: strangely they are not algae but flowering plants adapted for life in the sea. The Japanese weed which has recently appeared in the sea off Looe and elsewhere along the south coast could possibly pose a threat to the estuaries.

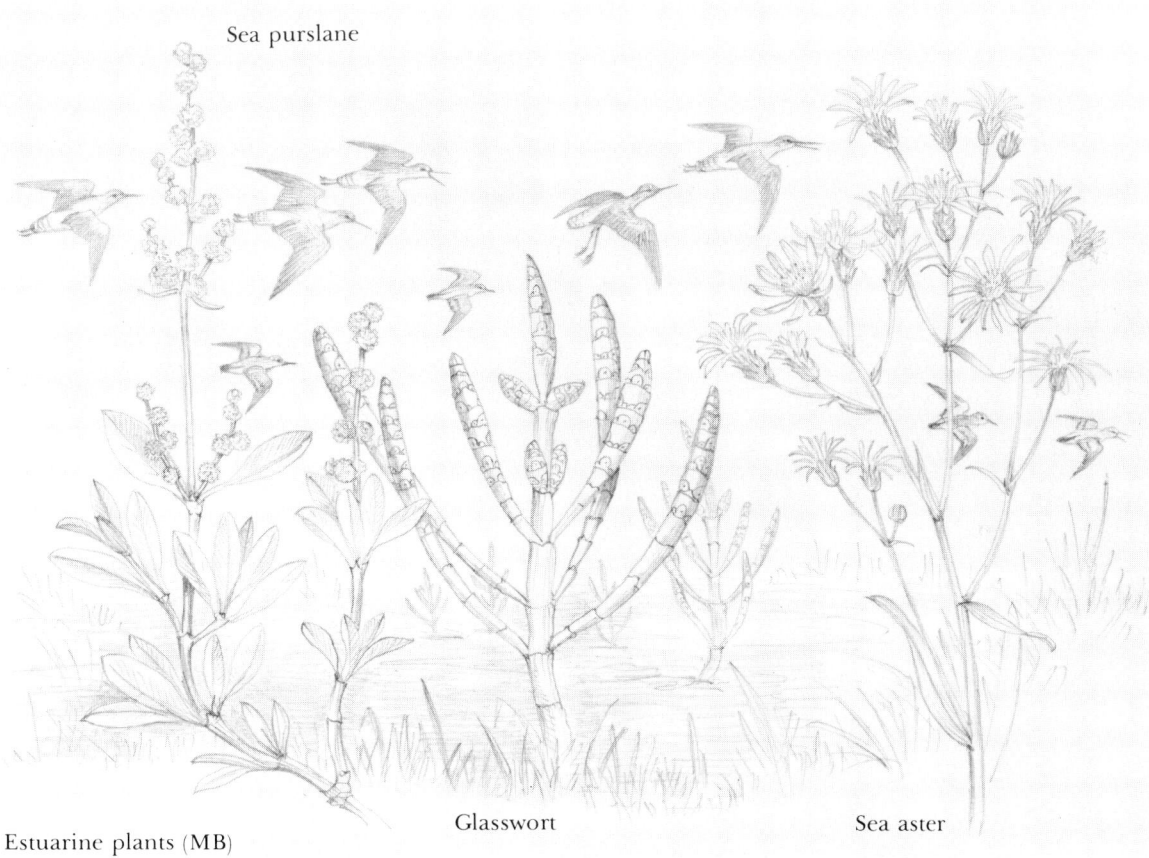

Sea purslane

Glasswort

Sea aster

Estuarine plants (MB)

The western shore of the Camel estuary is sheltered, particularly the coves and grassy slopes below Stepper Point, where conditions are ideal for many common butterflies. Two small subsidiary estuaries, Little Petherick and Pinson's creek where there is a heronry, add to the variety. The old Wadebridge-Padstow railway line, now converted to a footpath by the County Council, is an excellent place for observing birds on the mud-flats; it runs alongside the main estuary. The Camel is an important refuge for migrating and wintering waders and wildfowl, including occasional rare trans-Atlantic visitors such as the lesser yellowlegs and, once, the American stint. It is also visited by many other interesting birds. Ospreys appear most autumns. A peregrine occasionally hunts for sandpipers on the mud-flats. There are numerous resident breeding birds including shelduck whose young are preyed upon by gulls.

Near Trebetherick Point, on the opposite shore, there is a raised beach where fossiliferous Pleistocene shore deposits have been found. Elsewhere the shore is mainly sand with the unstable Rock dunes, which once buried St Enodoc church, and fixed dunes at Cant Hill where there is a quarry. The elegant wild maidenhair fern grows in crevices in the rock of nearby Porthilly Cove. Where the estuary begins to narrow, it is joined by the Amble river with its meadows and marshes, parts of which are tidal. These include Walmsley Sanctuary, of the Cornwall Bird-Watching and Preservation Society, which regularly attracts white-front geese, sometimes to be seen grazing among the sheep.

Dozmary Pool — cormorant flying over. (FC)

OPPOSITE ABOVE LEFT: The sand and shingle bar which separates Loe Pool from the sea, (CW/NT) and RIGHT: Loe Pool — the largest freshwater lake in Cornwall; (CW/NT) CENTRE: Dozmary Pool, on Bodmin Moor, and bordering marshland; (CGB) BELOW: Crowdy Reservoir overlooked by the rugged slopes of Rough Tor. (JB)

ABOVE: Old Tamar Lake — wildfowl sanctuary on the Devon border, (JB) and BELOW: Bude Canal; (JB)
OPPOSITE ABOVE LEFT: Respryn bridge near Lanhydrock — the Fowey in dry weather, (CGB) and RIGHT: low
water on a tidal reach of the River Fowey; (CGB) CENTRE LEFT: Salt marsh and tidal flats near Perranarworthal;
(CGB) RIGHT: Royal Fern — well distributed in Cornwall in suitable habitats, (CGB) and BELOW: the Lake on Red
Moor Reserve — willow carr in the background. (CGB)

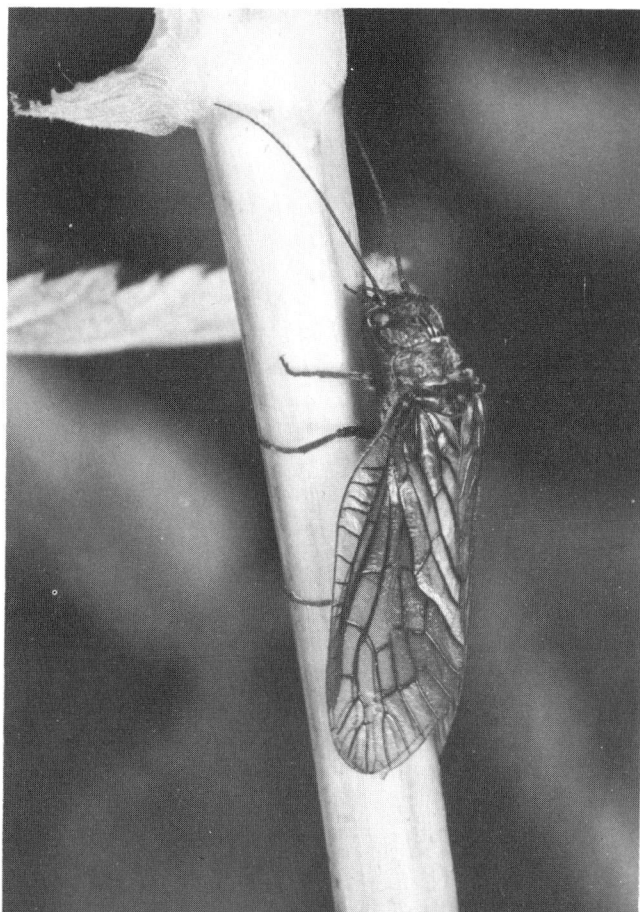

OPPOSITE: The common Tussock Sedge on damp ground at Boconnoc, (NCC) ABOVE LEFT: caterpillar of the Buff-tip moth on sallow — common on damp ground where sallows grow, (CGB) RIGHT: adult alder fly, frequent near lakes and reservoirs in Cornwall, (CGB) and BELOW: Bewick's swans on a pool in Marazion Marsh. (JB & SB)

ABOVE: Teal — sometimes present in large numbers in winter, (JB & SB) and BELOW: the familiar Dabchick, or Little Grebe, nests on many of Cornwall's wetlands. (JB & SB)

ABOVE LEFT: Tufted Duck — one of the commonest winter visitors to Cornish reservoirs; (JB & SB) RIGHT: Avocets regularly winter on the Tamar; (Michael Richards/RSPB) CENTRE LEFT: the Bar-tailed Godwit frequents mud flats near the coast, (JB & SB) RIGHT: Greenshank feeding in shallow water, (JB & SB) and BELOW: Greylag Geese — scarce passage migrants. (JB & SB)

ABOVE: Fringed water lily, (CGB) and BELOW: White water lily. (CGB)

Moorlands

Golden Plover and Lapwing above Davidstow Moor. (FC)

As well as the major granite areas there are small stretches of moorland, underlain by various geological formations, all over Cornwall. Many are undrained or partly drained farmland and are little more than rough pasture and willow carr. Some are true wetlands of which examles have been given in the last chapter. Others are on ground too steep or too broken for agriculture.

Bodmin Moor, a true granite moor, is the highest and most extensive of Cornwall's moorlands — even so, the altitude is not great enough to support a genuinely high-level (or alpine) flora. Bounded by its rivers (Inny, Lynher, Camel and Fowey in its westward flowing reaches) it is bisected by the A30 trunk road, which follows an ancient ridgeway track. The moor consists mainly of rounded hills and hog's back ridges with occasional granite outcrops and clitter slopes. Several of the higher hills are crowned by genuine tors, mostly serrated spines forming a series of miniature summits. The structure of the granite is such that it produces joints and lines of weakness which sometimes cut the rock into fascinating shapes, such as Corner Quoit in Cardinham and the famous Cheesewring above a granite-with-topaz quarry.

The highest and most rugged areas are the northern hills, where Brown Willy (1,377 feet) and Rough Tor (1,311 feet) dominate the neighbourhood, and the eastern block of several rocky summits — culminating in Kilmar (1,280 feet) — which stand around the marshy Twelve Men's Moor, so named for the prosaic reason that twelve men leased it from Launceston Priory in the Middle Ages. Several of the higher hills are virtually rock-free and there is much open rolling country with occasional aberrant outcrops such as Lanlavery Rock, Grey Mare and Elephant Rock below Hendra Beacon.

The most important source areas for moorland streams and rivers are the valley bogs around Rough Tor and Brown Willy; others rise close to the line of the A30. The Fowey flows southwards from Brown Willy, turns west after leaving the fine Lamelgate valley and takes in a number of tributary streams, including the Loveny, which is being dammed to become Colliford reservoir. Redhill marsh will be flooded, with the loss of some pale butterwort and other marsh plants, but two miles of the Loveny itself are destined to become nature reserve — a fairly narrow strip of water with botanically rich verges. Withey Brook, the finest upland stream in Cornwall, rises near Kilmar and flows down steeply wooded slopes above Trebartha to join the Lynher river. Most moorland rivers follow the same kind of course. Rising in marshes, they cross the moor as relatively placid streams with rocky or gravel bottoms, and pass through rough pastureland and more marshes wherever the water is inclined to overflow its banks. As they leave the moor, they plunge down small rapids and enter wooded reaches on their way to becoming part of lowland Cornwall.

Except for willows beside streams and occasional stunted hawthorns and rowans, the high moor is virtually treeless, except where there are modern conifer plantations. There are bogs and marshes in every valley but Dozmary is the only lake. Pools are numerous, particularly in areas formerly worked for china clay: Temple Tor pools, for example, which are much appreciated by birds. The construction of reservoirs has greatly increased the amount of standing water, thereby adding a totally new dimension to the ecology of the moor.

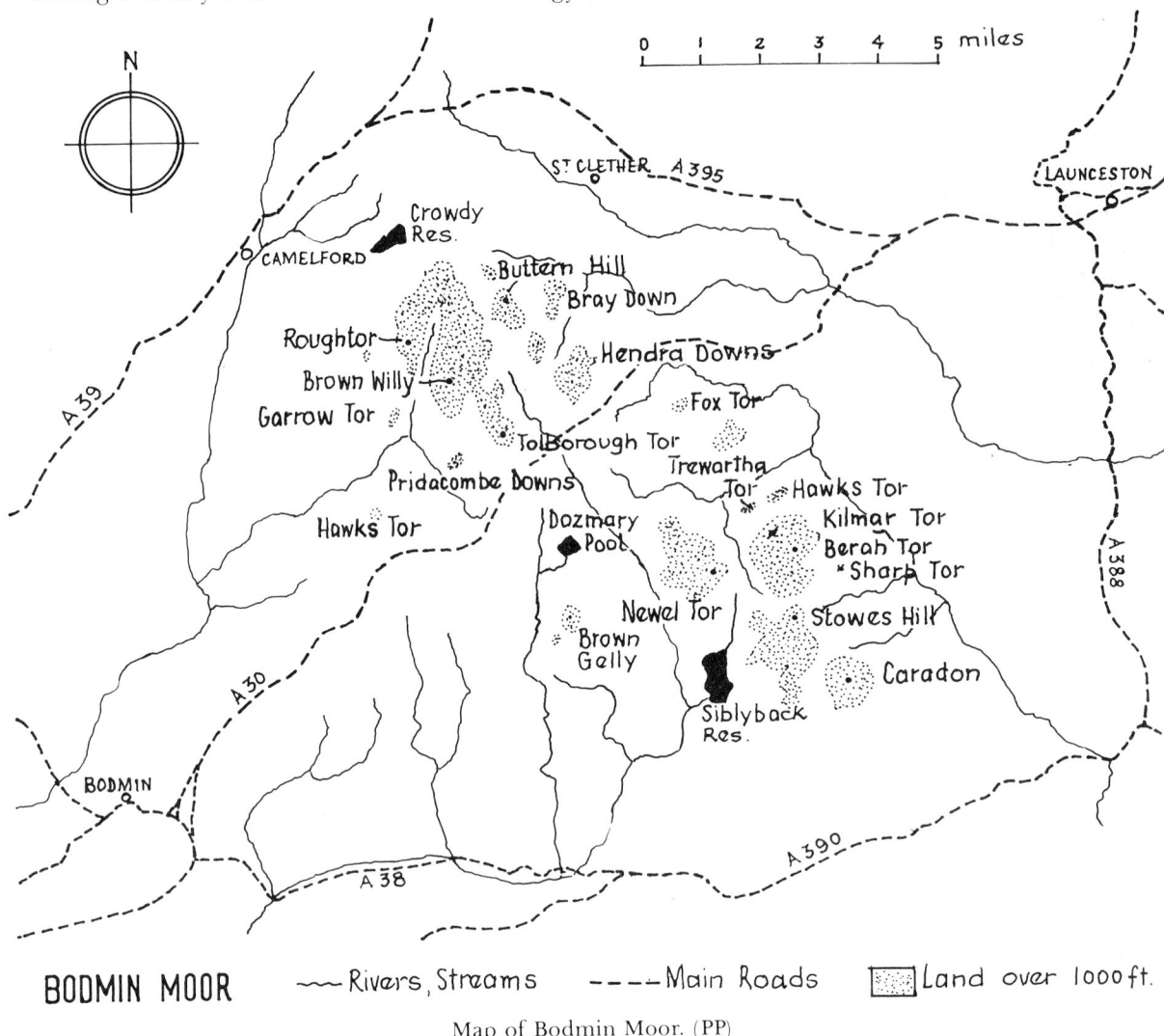

Map of Bodmin Moor. (PP)

During the period following the last Ice Age, Bodmin Moor seems to have been at least partly wooded with dwarf birch, and least willow in upland valleys, stunted junipers with crowberry and heathers on exposed hillsides, and oak in valleys at the edge of the moor. Man soon began to clear the land; soils, already poor because of inadequate drainage through the granite, deteriorated further and became increasingly acid. There are signs everywhere of early man's activities with stone circles, hut settlements, barrows, Stone Age flints (found near Dozmary) and the chamber-tomb known as Trethevy Quoit. There are ancient crosses and holy wells from the early centuries of the Christian era and frequent signs of former mining, china clay extraction and quarrying — an old quarry near Bowithick is now occupied by both badgers and rabbits. The changing pattern of farming has resulted in abandoned farmsteads and derelict farm buildings, though the moor is still grazed throughout under control of a strong commoners' association. Some of the old stone walls are believed to date from the Bronze Age, when the first enclosures were made. Enclosures have continued ever since; the extent of unenclosed moor, now estimated at 29,640 acres, has been halved since 1938.

Bodmin Moor may be looked upon as including eight major habitats.

Open Moorland covers most of the area which is now largely grassland, dominated by fescue and bent grasses in drier areas and purple moor-grass in wetter parts. Tormentil, heath bedstraw, eyebright, lousewort, common milkwort and pale heath violet are common in fescue-bent areas

Moorland plants — Tormentil, Heath Bedstraw, Eyebright, Lousewort, Common Milk Wort. (MB)

but tend to decrease where exotic grasses have been planted. Where wetter areas are not too heavily grazed, heathers (ling, bell heather, cross-leaved heath) are still plentiful, often in association with common or western gorse; deer-grass (a sedge) grows in damp peaty hollows. Grassland at the edge of the moor is often invaded by bracken. Sheep and rabbits, neither of which eat bracken, encourage this tendency. Cattle, which both nibble and trample the young fronds, have the reverse effect, so that the incidence of bracken varies with farming practice. Hawthorn, willow, elder and brambles also invade these peripheral areas if allowed to do so.

The open moor provides the main inland habitat in Cornwall for such birds as meadow pipit — and the cuckoo which parasitizes it — skylark, wheatear and linnet; whinchats also occur but are not numerous. Birds of passage include large flocks of lapwing and golden plover. The common green and meadow grasshoppers are the commonest of their kind; the mottled grasshopper is more local. Several butterflies occur: marbled white and speckled wood are both fairly common; dark green fritillary, which associates with violets, and green hairstreak, which feeds on gorse and bilberries, are worth looking for.

Tors and Clitter Slopes are, for many, the most typical features of the moor. The rocks and boulders tend to create small pockets of land protected from grazing animals, with the result that there are more frequent heathers and greater diversity in the vegetation. Gorse is often plentiful as on Brown Willy. There are colourful patches of English stonecrop, the starry white flowers tinged with red. Bilberries are common but true 'bilberry moor' *(or Vaccinium)* is restricted to parts of Hill

English stonecrop

Bilberries (MB)

Tor in St Cleer. Occasional stunted hawthorn and rowan trees grow on sheltered slopes with, but rarely, holly and oak. *Juncus* rushes occur in damp flushes. Ferns are plentiful in cracks and crevices in the granite, among them the Tunbridge and Wilson's filmy ferns; the rare beech fern used to grow on the highest tors but has not been seen for some years. Certain mosses grow in these exposed places; lichens include the nationally rare beard lichen, *Usnea articulata*.

Some of the more massive rocky summits are used by nesting ravens and kestrels but the buzzards seen soaring above the heights nest in woods lower down. The wren is probably the commonest small bird of the rocks. Some clitter slopes have been colonised by foxes, others by badgers. The common field grasshopper frequents the tors, and the mountain or bilberry bumble bee, with tail and most of the abdomen yellowish-red, may survive on the higher tors where there are heathers and bilberry, but is more frequent on higher ranges in other parts of the country. The grayling is probably the commonest butterfly but is remarkably difficult to see; burnet moths have also been observed in this habitat.

The Valley Bogs and Marshes have suffered greatly from drainage but still support an interesting acid-bog flora, with such specialised plants as common cotton-grass and harestail. Marsh violet, marsh speedwell, marsh St John's wort and bog asphodel are plentiful. Heath spotted orchids are

Tunbridge filmy fern.

Wilsons filmy fern. (MB)

84

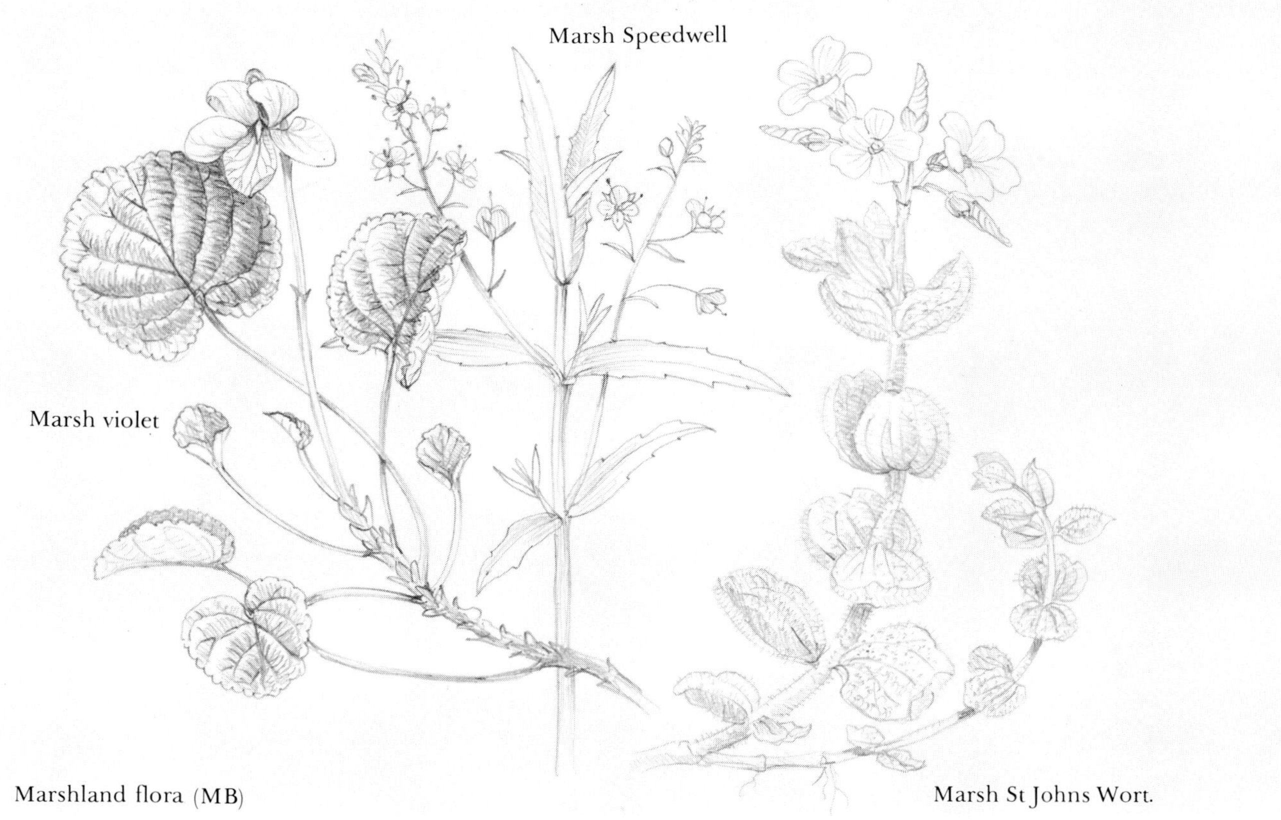

Marsh Speedwell

Marsh violet

Marshland flora (MB)

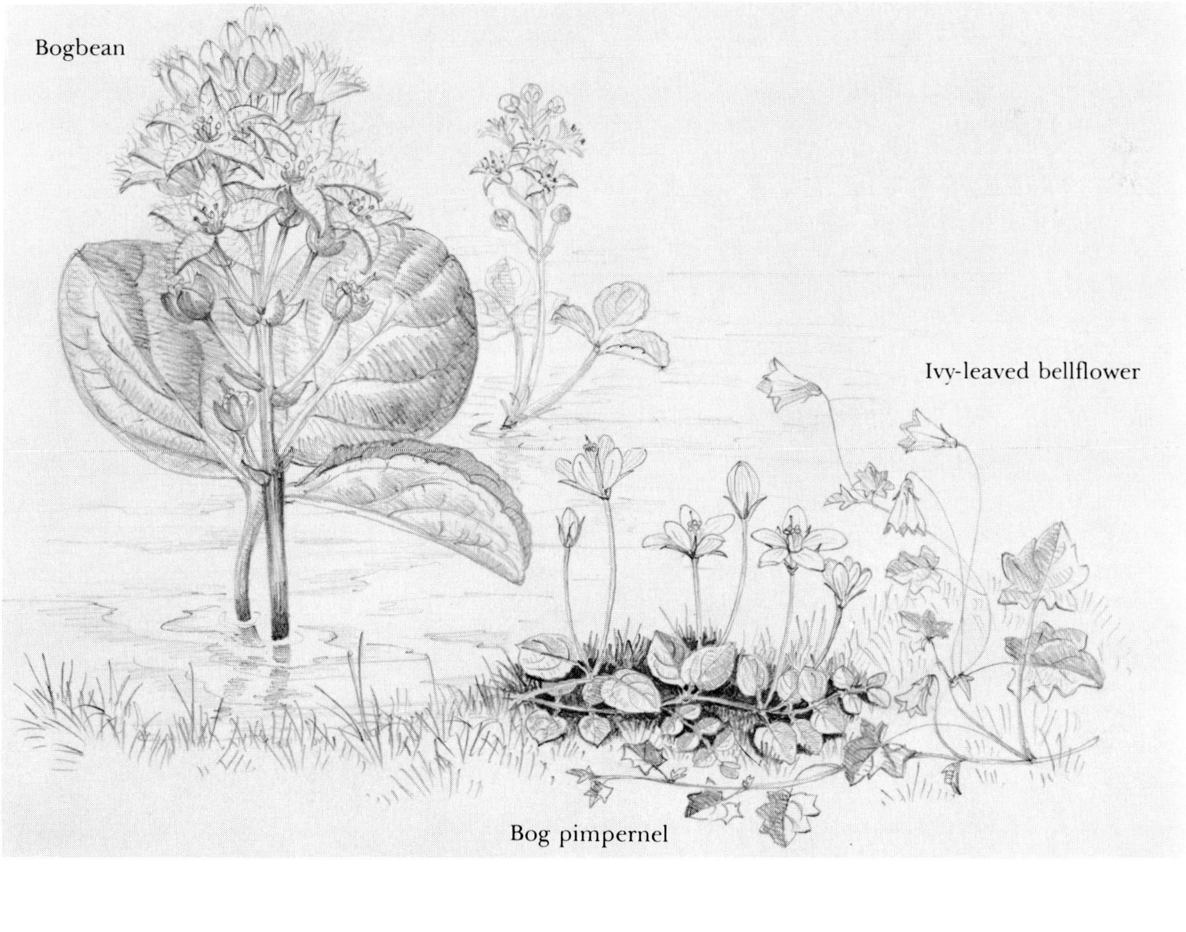

Marsh St Johns Wort.

Bogbean

Ivy-leaved bellflower

Bog pimpernel

common and the rare bog orchid occurs in a few places. Insect-catching plants are represented by pale butterwort and two species of sundew. There are numerous sedges with white beak-sedge in a few places; heathers and gorse grow on drier margins. There are few extensive carpets of *Sphagnum,* the true bog moss, but small patches are frequent — among them the rare red-tinged *S. magellanicum* which occurs in Bowithick bog but nowhere else in Cornwall. Hair-moss is one of the commoner species. In the wetter parts of many bogs the pink-tinged flowers of bogbean stand out from a mass of leaves. In valley bottoms, purple moor-grass tends to grow in tussocks with deer-grass and several kinds of rush. Mats of bog pimpernel are common in damp turf.

Virtually every valley on the moor is boggy, though details of the vegetation vary. In the small portion of Crowdy marsh which remains unflooded by the reservoir, some seventy flowering plants have been recorded, including bog orchid. Little Care marsh on Cardinham Moor is a wet valley with abundant bogbean, bog asphodel and ivy-leaved bellflower. In the Warleggan valley, where there are many old clay pits, sundews are particularly abundant. The marsh below Dozmary Pool is another area of great interest.

These valleys provide the only nesting sites in Cornwall of snipe, redshank and dunlin; the latter has bred a few times near Dozmary. They are the principal breeding places of curlew, usually in isolated pairs. A few pairs of mallard nest, and teal breed beside Temple Tor pools. The common frog is reasonably plentiful; the marsh frog, an exotic introduced species, is present in a few bogs. Palmate newts occur in both bogs and streams. Adders may sometimes be seen curled up on tussocks of moor-grass enjoying the warm sun — grass snakes are more frequent away from the moor. The most commonly seen dragonflies are the broad-bodied libellula, the common sympetrum and the grey-green common coenagrion, a damselfly. The large marsh grasshopper, which associates with bog asphodel, and the bog bush-cricket are widespread but nowhere common; the latter seeks out the wettest parts of bogs and places where cross-leaved heath occurs. The marsh fritillary butterfly is locally common. The moss carder bee, one of the species which makes wax pockets for storing pollen, has been recorded on Harpur's Down marsh and may have its main Cornish habitat in such wet moorland valleys.

The Still Water Habitat was described in the last chapter, and there is little more that need be added. The pools, lakes and reservoirs of the moor are inseperably linked ecologically with the adjacent marshland: notably so at Dozmary and Crowdy, natural and man-made lakes respectively. The most obvious still water plants in small moorland pools, some of which are almost completely covered by duckweed, are flote-grass, water crowfoot and water starwort. Rushes and sedges fringe the borders.

The Running Water Habitat is best thought of in conjunction with marginal vegetation and mid-stream boulders. The quieter reaches of moorland streams are usually richer in plant life than torrents, with such species as pondweed, floating spike-rush, also found in pools, whose leaves spread out in the current, lesser spearwort and mats of ivy-leaved bellflower. Cornish moneywort and heath spotted orchids grow on banks, sometimes in association with Tunbridge filmy and mountain buckler ferns. The splendid royal fern occurs below the level of the open moor.

Animal life is limited to species which can maintain themselves in the fast-flowing flood-waters which follow heavy rain. The dipper is a bird perfectly adapted to these conditions, and may be seen bobbing up and down upon a boulder or flying rapidly over the water before landing and then plunging in. Kingfishers occur only where the water flows more gently. Several dragonflies are common, notably the golden-ringed hawker, large red damselfly and deep blue demoiselle agrion. Most of the fish are small and local but trout, miller's thumb, dace, stone loach, minnow, gudgeon and brook lamprey all occur — some of these are also present in still water; grayling and trout have been introduced, the latter to augment the wild population.

Cornish Hedges, Stone Walls, Quarries and Derelict Buildings present a variety of artificial micro-habitats which may conveniently be lumped together. Hedges in exposed situations support such plants as mouse-ear hawkweed, wall pennywort and English stonecrop. In more sheltered localities there may be a plentiful covering of foxgloves, golden-rod and betony, with sometimes a few hawthorn or rowan trees. Devilsbit scabious, foodplant of the marsh fritillary butterfly, and sheepsbit grow near or at ground level. Old quarries, isolated rocks and some hedges give protection to gorse, heathers and — much less frequently — the greater butterfly orchid. Rock spurrey and the minute fingered saxifrage grow on walls and bare dry places. Ferns are frequent in this group of habitats: hard, hay-scented buckler and soft-shield ferns and lanceolate spleenwort.

Holes in hedges and old walls serve as refuges for small mammals and nesting sites for such birds as wrens and wheatears. Derelict buildings are often taken over by robins, blackbirds, swallows, barn owls and other common species. During recent years, the common redstart has taken to breeding in old walls on the moor and is becoming more numerous in Cornwall as a result. Ravens, jackdaws and kestrels sometimes nest in old quarries, while the steadily diminishing number of nightjars breed on one or two derelict mine tips at the edge of the moor.

Deciduous Woodland occurs in sheltered valleys on the margins of the moor but not to the north-west, where high ground almost reaches the Atlantic. Once the first clearings had been made, the uplands seem to have remained largely treeless. Only one wood actually on the granite was recorded in the Domesday survey: at Halvana in the valley below East Moor where there is now a plantation. Those woods of the moorland verges which have not been replanted are mostly oak and ash woods of the Draynes type. They bring a distinct fauna to the edge of the moor with green and great-spotted woodpeckers, nuthatch, treecreeper and tawny owl. Buzzards, which nest in the woods, hunt over the open moor. Warblers breed during the summer. Badgers and other woodland mammals are plentiful. Red deer from woods beside the Camel occasionally visit the open moor.

Conifer Plantations fall into two categories: those which have replaced old deciduous valley woods, and those planted on the open moor. The valley plantations are mainly in the Glynn valley and around Cardinham and are part of the changing pattern of woodland management. Plantations on the open moor introduce a completely new habitat; provided the area planted is not excessive and that no drainage of marshes is involved, this is not necessarily detrimental. A modest plantation can enhance the moor, as has happened at Davidstow, where the conifers help to obliterate an old wartime aerodrome. Other plantations have been established on Smallacoombe Downs and on Rough Tor, as well as at Halvana, to give a total area of 2,220 acres on the open moor — additionally 1,730 acres of deciduous valley woods have been replaced by plantations. At different stages of growth with rides and clearings, these plantations have brought to the moor new conditions which are proving attractive to both vertebrate and invertebrate animals. There are now two small heronries in moorland plantations. They have also been responsible for the first-ever Cornish breeding records of the lesser redpoll, which now nests successfully in both Davidstow and Smallacoombe plantations. Long-eared owl, firecrest and goldcrest and jay roost and occasionally nest in them. They provide winter roosts for millions of starlings.

There are many other moorland areas, but Bodmin Moor is not only by far the largest, its scale is such that the major habitats are easily defined. Of the remaining granite areas (other than Penwith which has already been described) Hensbarrow (1,027 feet) has been so extensively developed for china clay that little natural moorland remains. Carmenellis has been much worked over for tin and other metals and — except in former mining areas which have their own special interest — is now largely an area of gentle hills given over to farming. The moorland was greatly reduced during the 19th century when many miners, forced to supplement the inadequate wages paid by a declining industry, took to smallholdings. There still are interesting pockets of moorland, though most have been much modified by past mining or farming. Porkellis Moor, near Helston, is a good example. It has considerable diversity of habitat with heathland, rough grazing, scrub, willow carr, several ponds and an area invaded by bracken. There is nothing unusual about either the flora or the fauna except for the large number of different dragonflies and damselflies. Parts of Porkellis seem likely to be redeveloped for alluvial tin mining or to be drained for agriculture. Among interesting plants found on some of these moors is the now rare Dorset heath which occurs on Silverwell Moor, Newlyn Downs and one or two other places, as well as on Ventongimps and the Lizard.

At the northern edge of the Hensbarrow granite there are rocky outcrops rising above the killas as at Belowda Beacon, Castle Downs and Roche. The distinctive point of Helman Tor, on the eastern edge of the granite, overlooks Red Moor reserve. There is also Goss Moor, another area threatened by development. Immediately to the south of the A30, it is mainly an area of rough grazing dominated by moor-grass and harestail, but there are pools with yellow water-lilies, damp patches, incipient woodland and old mine buildings. Pale butterwort, lesser bladderwort, wavy St John's wort and marsh violet occur. Both lapwing and curlew breed on Goss Moor.

Plants of moorland ponds (MB)

Lesser Bladderwort

Yellow waterlily

There are several heathland areas, both wet and dry, on the culm measures as well as on the killas; most are at least partly reclaimed. Wrasford Moor, near Kilkampton, is the most northerly moor in Cornwall and the home of the uncommon marsh thistle. Laneast Moor is in part damp heath, with pale butterwort and a few fragrant orchid plants, and in part dry grassland with some heathers on a peaty soil. Tregeare Down, also near Launceston, is heathland on a chert ridge with willow carr and some tussock sedge. There is dry heathland on higher ground in many parts of Cornwall: St Breock Downs, for example, where Montagu's harriers once nested, but which is now almost entirely reclaimed for agriculture where not planted with trees. There are steep-sided valleys throughout the county where heathland characteristics are evident, wet valleys with willows, and small areas of marsh vegetation. Many show signs of past tin-streaming operations which would have destroyed any extensive *Sphagnum* bogs that may once have existed. There are also narrow fringes of heathland along many parts of the coast, between cultivated land and the cliff-edge, with superb seasonal displays of colourful gorse and heathers.

Buzzard in flight above a Bodmin Moor Tor. (FC)

ABOVE: Most of Bodmin Moor is open grassland used for grazing; (JB)
LEFT: hills capped by true tors — Rough Tor, (JB) and RIGHT: the rocks
and clitter slopes of Bodmin Moor. (JB)

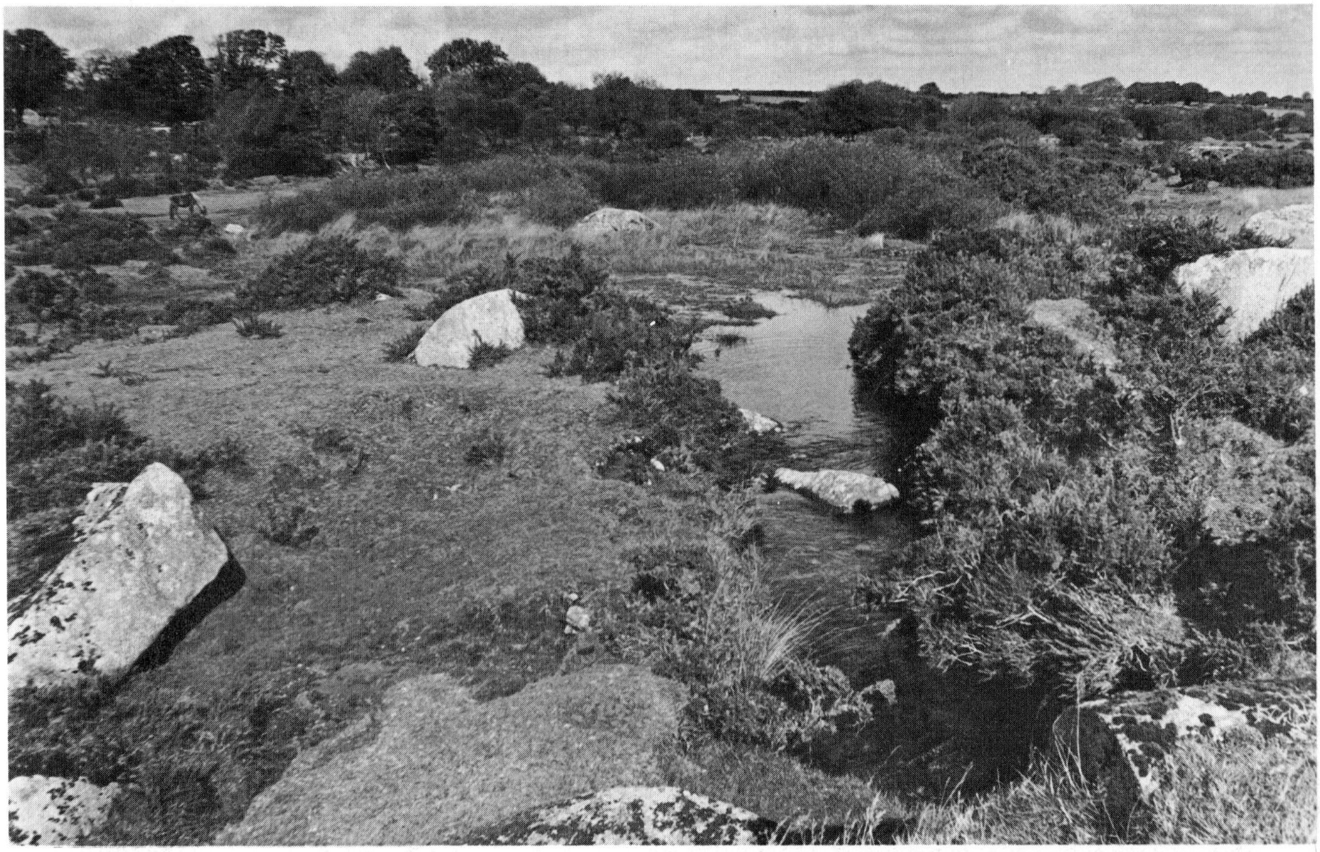

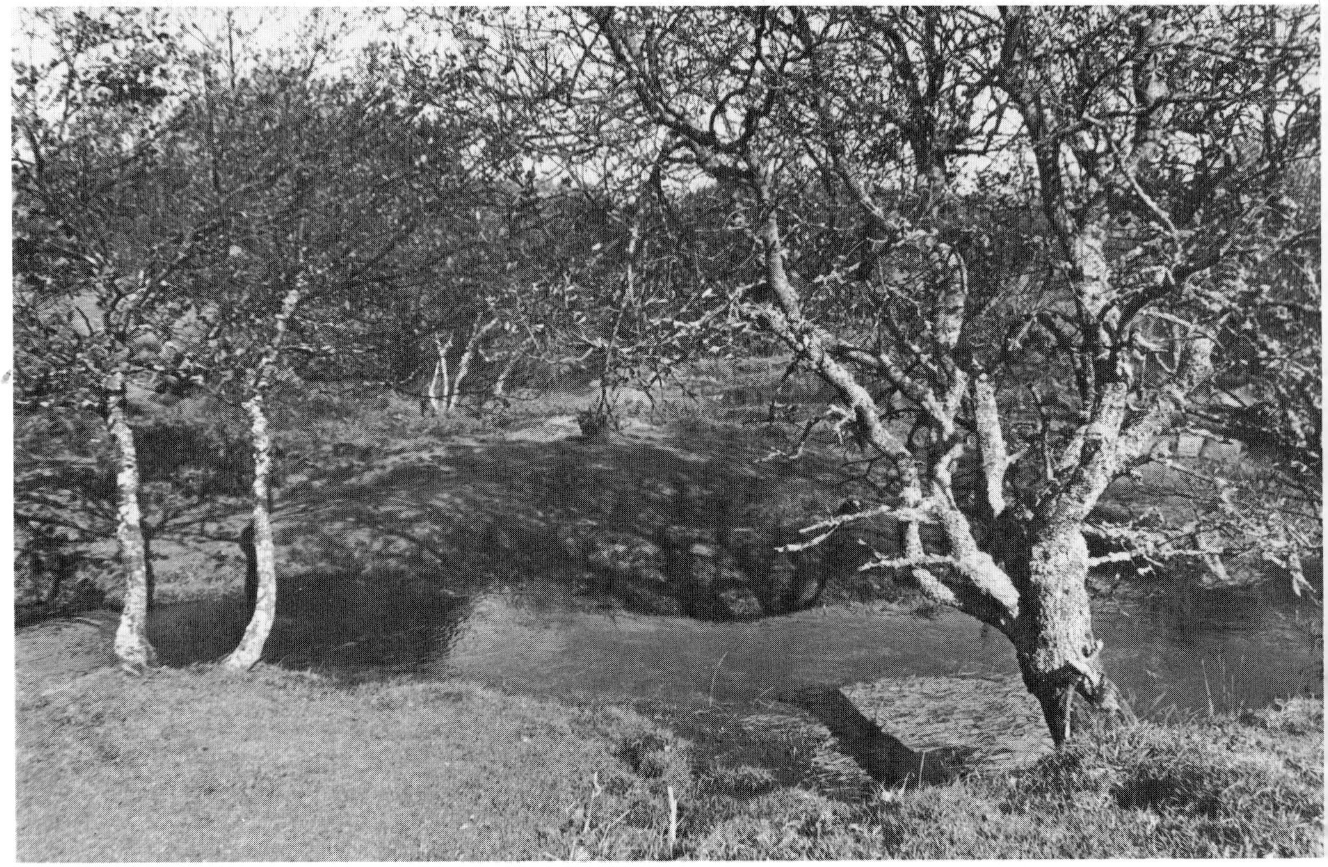

OPPOSITE ABOVE: Bowithick Bog — site of a rare moss and many other marsh plants, (JB) and BELOW: stunted trees beside streams and in sheltered hollows. (JB) ABOVE: Breney Common and Helman Tor — land recolonised by vegetation after opencast mining; (CGB) CENTRE: Cotton grass, widespread on wet heaths and moors; (CGB) LEFT: Saxifrage, Ivy and Lichens on rock near Roche, (CGB) and RIGHT: the harmless grass snake is quite common in damp areas on the moors. (CGB)

LEFT: The Tiger beetle with prey — favours dry heaths and moors; (CGB)
CENTRE: the Robber Fly occurs in dry moorland areas — lays its eggs in
dung; (CGB) RIGHT: Meadow grasshopper — common on the open Moor,
(CGB) and BELOW: Ravens nest on certain rocky summits of Bodmin
Moor. (JB & SB)

ABOVE LEFT: Whinchat — a few pairs breed on the higher moors, (JB & SB) and RIGHT: huge flocks of Lapwings are often seen on Bodmin Moor in winter; a few nest there; (JB & SB) CENTRE: Dippers favour the running water of moorland streams; (JB & SB) BELOW LEFT: Meadow Pipit, the commonest small bird on the open moor, (JB & SB) and RIGHT: Skylark on a granite post, encrusted by lichens. (JB & SB) OVER: An upland stream — Penpont Water. (JB)

Woodlands

Carclew beechwood as it was in the 1940s, now cut down. (FC)

When ice finally receded from Britain around 8000 BC, the first tree to invade the tundra of Cornwall was birch, followed by Cornish elm, alder, hazel and sessile oak, with willows on damp ground. Oak had become the dominant wildwood tree before man began to clear the land for his crops and domestic animals. During the Iron Age (350 BC to AD 50), oak was much in demand for smelting tin for sale to Phoenician merchants from the Mediterranean. This trade was developed and extended under the Romans. They not only used oak logs for smelting tin and burning charcoal, they used the bark for tanning, a small demand for which continued until the the 1970s. The Romans also introduced coppicing (periodic cutting for small timber) on a seventeen year rotation, working Merthen Wood, a sessile oak and ash wood beside the Helford River, and Lantyn Wood by the River Fowey in this way. Coppicing was later adopted throughout Cornwall and remained the normal practice in woodland management until late in the 19th century when improved methods of distributing coal reduced the demand for wood-fuel. Many old coppiced woods still stand, direct descendants of the wildwood, the nearest approach there is to natural woodland in Cornwall today.

Domesday survey (1086) listed a large number of woods in Cornwall, mostly in the south of the county and the hinterland of the Stratton Hundred. Elsewhere few woods were recorded and few now exist except in deep valleys. During medieval times, the landscape became stabilised and much of it would be recognisable to us today. There were woods producing timber trees, woods producing 'underwood' (ie coppice mainly used for firewood, fencing and other country trades) and wood pasture. The two former were often combined with timber trees left standing in the coppice. There was also scrub and 'waste', with small trees including willows used for basket-making, which provided ordinary people with many of their domestic needs. During the 13th century 'the men of Cornwall' purchased the rights of disafforestation from King John and though mainly to do with hunting, this meant some large scale clearing of woodland. The Black Death (1348) reduced the human population and indirectly allowed a fair amount of reclaimed land to revert to scrub and woodland.

Throughout most of the recorded past, woods were cut without destruction, trees recreating themselves from their own roots as when coppiced — while coppicing is thus a form of active conservation it also results in slow decline over a long period. There is no evidence that ship-building, one of the main uses of timber in Cornwall, actually destroyed woods or that Royalist land-owners did so when selling timber to pay their fines after the Civil War, though this is often said. It is more likely that they kept their trees as a reserve which could be turned into cash in emergencies, instead of grubbing them out to produce farm-land.

Many kinds of tree have been introduced down the centuries, to become an accepted part of the county's woodland. Sweet chestnut, never to become a particularly common tree in Cornwall, was brought in by the Romans and sycamore not long after Domesday, though most of the now familiar introductions were planted during the 18th century or later, as Cornwall's climatic suitability for growing exotic species came to be appreciated: Scot's pine, silver fir, Douglas fir, western hemlock, western red cedar and holm oak. Most of the beech seems to have been planted in the 18th century. However, fossil evidence indicates that beech was once a native tree in Cornwall but that it died out. There are many fine collections of both native and exotic trees; most are in private grounds occasionally or regularly opened to the public (Pencarrow, near Bodmin, for instance) but there are several National Trust parks and gardens (notably Glendurgan, Lanhydrock and Trelissick) and an extraordinary collection of southern hemisphere trees at Tresco Abbey in the Isles of Scilly.

The term 'pasture woods' refers to parkland or wooded farmland used mainly for grazing. Once common all over the county, these are now largely restricted to a few parks such as Trelissick, Werrington, Mount Edgecumbe, Tregothnan, Bocconoc and Lanhydrock where there is also a fine stretch of deciduous woodland. Ancient parks and undisturbed woods are particularly rich in epiphytic lichens, fungi and the invertebrate fauna of dead wood. Bocconoc, a 1,500 acre estate near Lostwithiel, supports one of the best lichen floras in western Europe. It is a complex of closed canopy woodland (mostly oak and beech), park and ancient pasture woodland on steeply sloping ground. There are 180 lichen species, including *Porina hibernica* not known anywhere else in Britain, and others found in one or two localities only, as well as the tree lungwort. For lichens to develop on this scale the atmosphere must be humid and unpolluted, the light intensity high and human interference minimal over a long period.

Although a number of old deciduous woods have been replanted with alien conifers, many remain in different parts of the county. A majority of these are valley woods, some too small for economical re-afforestation or on ground too steep or too waterlogged; they are among the most important natural habitats in Cornwall. An outstanding example is Draynes wood beside the River Fowey, where it plunges down the Golitha Falls on leaving the high ground of Bodmin Moor. It is a damp oak and ash wood on steep slopes with dripping rock faces. The dominant trees are both common and sessile oak with ash, hazel, beech, holly and sycamore. There are numerous mosses and liverworts, several kinds of fern including, near the river, the little Tunbridge filmy fern. The ground flora is particularly rich and varied with sanicle, wood-sorrel, bilberry and honeysuckle. These and many other colourful wild flowers are frequent in Cornish valley woods: snowdrop, bluebell, wild daffodil, wood anemone which is almost confined to ancient primary woodland, golden saxifrage, the delicate little moschatel, the prostrate mat-forming Cornish moneywort and occasionally the rare Cornish bladder-seed. Such woods are a superb sight in spring and early summer.

The Trust has several woodland reserves. One of these, Peter's Wood in the Valency valley, is leased from the National Trust; it is a remnant of primeval woodland which has been little interfered with except by coppicing. On steep hill-side, with a footpath along its eastern edge and a stream below, the canopy trees fall into three distinct groups: at the top of the slope, sessile oak with polypody and broad buckler fern growing in the forks of trees; beech, birch, sycamore and common oak in the centre; birch, rowan and willow at the lowest level near the stream. As in most Cornish woods, hawthorn, holly, hazel and ash are the main under-storey trees with ferns, wild roses and herb robert among the shrubs and ground flora. Wet patches are brightened up by hemp agrimony, red campion, meadow-sweet and yellow flag. The wood supports an exceptionally high population of snails and slugs, probably because it is underlain by the Barras Nose beds, which are volcanic intrusions containing limestone, thus providing the calcium needed by many molluscs for their shells.

Hawke's Wood reserve near Wadebridge, of which the Trust owns the freehold, is in many respects similar: an old sessile oakwood with a variety of other species in both canopy and shrub layers and a wide range of flowering plants and ferns. There is an abandoned quarry with wet slate

Woodland flora (MB)

Bluebell

Daffodil

Moschatel

Cornish Moneywort

Wood Anemone

Snowdrop

faces and a single alder buckthorn tree growing above it. Unlike Peter's wood where few birds nest, Hawke's wood provides for an excellent variety of woodland birds including buzzard, jay, tawny owl, green and spotted woodpeckers and tree creeper. A pair of dippers nest on the bank of a stream which bounds part of the wood. Bats of unidentified species (possibly noctule which roosts in hollow trees in Cornwall) are commonly seen in the twilight; palmate newts breed in a small pond, and six species of bumble bees have been recorded, mostly visiting cow-wheat.

Hawthorn
Holly
Hazel
and Ash

(MB)

Pelyn, a different kind of wood, is divided into two sections, the smaller of which consists mainly of sessile oak, sweet chestnut and sycamore around an outcrop of rock with maidenhair spleenwort growing in fissures and a badger sett below. The main block (80 acres) has much the same canopy but with more ash, some fine beech trees and a few conifers; it was probably a typical oak-ash wood into which beech was planted during the 18th and 19th centuries. When the Trust took over responsibility laurels had proliferated in the valley, where there is a pleasant stream, smothering everything except large trees. Management has been concentrated on clearing away the laurels and letting light into the wood to encourage growth of woodland plants. Numerous ferns grow in damp areas. Woodland birds and mammals, including badgers, are plentiful.

Pendarves, near Camborne, is managed by the Trust in agreement with the Forestry Commission, who are the owners. It is a mixed wood in a sheltered situation with a high water-table, caused by a stream and a lake of some four acres, which gives the reserve its peculiar character. Originally mixed woodland with oak, beech, alder, lime, which is rare as a woodland tree in Cornwall, and willows by the lake, it has been partly replanted with conifers, with the result that there is an unusually wide range of habitats within an area of some 55 acres. Birds range from woodpeckers, tawny owl and spotted flycatcher to moorhen, little grebe and coot; ten species of dragonfly have been recorded. Pendarves is being opened up for educational purposes and provided with a nature trail.

Luckett reserve is a small mixed woodland section of the Duchy of Cornwall's Greenscombe valley plantation complex, an outstanding example of imaginative planting of hardwood trees and conifers in combination. The reserve itself has evolved from abandoned strawberry beds and orchards, natural woodland with fine trees in an old hedge and a corner of moorland. As well as supporting a good range of plants, including Cornish bladder-seed and three species of orchid, it attracts numerous butterflies and other insects. Among the butterflies found in the reserve and its immediate environs are six of the ten British fritillaries; one of these is the rare heath fritillary, which here feeds largely on ribwort-plantain, not cow-wheat as is usual elsewhere. The reserve itself (10 acres) is not large enough to provide for a viable population of this butterfly, but is being successfully managed to allow for a breeding reservoir, chiefly by clearing undergrowth and brambles to encourage ribwort-plantain. The Duchy of Cornwall has recently bought a small area of rough pasture and gorse adjoining the reserve to provide more space for the fritillary.

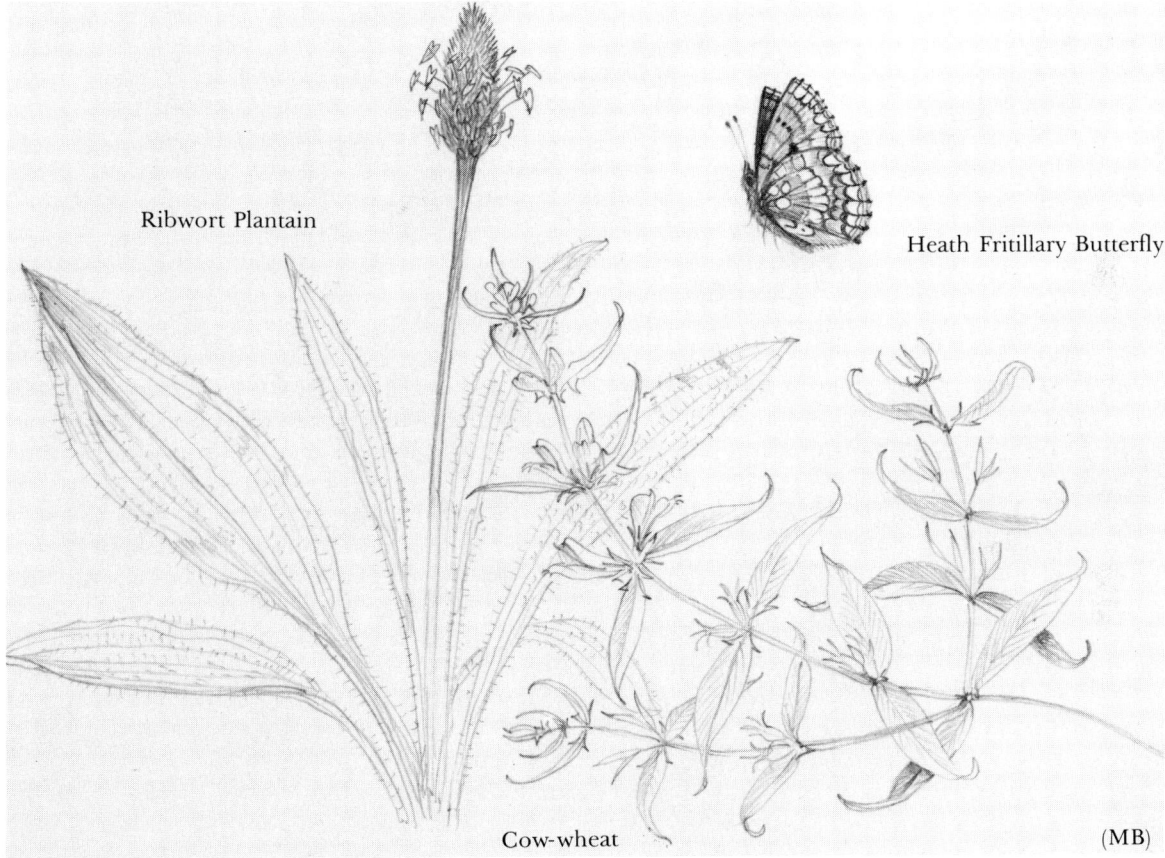

Ribwort Plantain

Heath Fritillary Butterfly

Cow-wheat   (MB)

With Draynes and the other broad-leaved woods that have been briefly touched upon, these reserves form an interesting series of woodlands in transition from natural descendants of the wildwood to modern plantations. Except for some of the more obvious birds and one butterfly, little has yet been said about the fauna which is of outstanding interest in old woods, the reservoir of so much of the country's traditional wildlife, the homeland of many common mammals and of a myriad of invertebrates. Woodland birds are less numerous than in many southern counties as Cornwall, particularly in the far west, is not heavily wooded. The result is a westward decrease in some of the tits and warblers and of such birds as nuthatch and tree-creeper. There is compensation in the widespread furze and scrub and the hedges to which some woodland species have adapted themselves — green woodpeckers have even started breeding in scrubby willow carr in some coastal areas.

Badgers, foxes and squirrels are the most obvious woodland mammals, but the first two are now more closely associated with the agricultural scene. It is just possible that the red squirrel still survives in west Cornwall, where it lasted longer than in most of England, but the last fully authenticated sighting seems to have been in the early 1970s in Treva Croft wood near St Ives; meanwhile the grey has spread throughout the county. The long-tailed field mouse is the commonest small rodent of Cornish woodland where it feeds on seeds, seedlings, buds, nuts, bark, snails, insect larvae and grass — a catholic diet which accounts for its success. Bank voles and shrews, which are insectivorous, are also numerous, but the attractive little dormouse has become extremely uncommon. How woodland bats are actually distributed in Cornwall is largely unknown: the noctule, a large bat with small ears, is probably the commonest species; Leisler's and the barbastelle have only been recorded once in recent years, and the pipistrelle is more often found in buildings.

The heath fritillary is not the only woodland butterfly which is becoming rare. Both high-brown and pearl-bordered fritillaries are distinctly uncommon, but the silver-washed, which feeds mainly on violets, is still well distributed in east Cornwall, as is the speckled wood throughout the county. The jagged-winged comma is nowhere common but sometimes visits gardens. The status of the purple hairstreak is difficult to assess as it is restricted to oakwoods and usually flies around the woodland canopy. Woodland moths are numerous but mainly nocturnal. The bumble bees most likely to be seen are the large earth bumble bee and *Bombus lapidarius,* a black bee with a red tail which lives in large colonies and is much attracted by the flowers of sycamore. There are grasshoppers and bush-crickets in deciduous woods, including that splendid insect the great green bush-cricket which, however, is more often to be seen in shrubby areas on the cliffs. The small pale oak bush-cricket is much more likely to be overlooked. Spiders hunt along the branches of trees; the greenish coloured crab spider is commoner in oak scrub than on mature trees; ground beetles, earwigs and woodlice search leaf-litter for food. There are several species of ant including *Lasius fuliginosus,* a shiny black ant which forages in trees and nests in rotting stumps, and the wood ant. Both these are worth looking for as neither is considered common in Cornwall.

Old woods and the marvellously varied wildlife that survives in them deserve to be preserved, and conservationists should do their utmost to save them where they can. But this is a period of transition following long years of neglect in many if not most broad-leaved woods, and they cannot all be retained. Woods and forests have always been put to the service of man; it is no use imagining that they can be left untended and undeveloped or managed according to the methods of past centuries. In Cornwall the aim must be to preserve a sufficient and representative sample of different kinds of wood in different parts of the county. Fortunately there are major woods safely in the hands of the National Trust, and the extensive Marsland valley woods on the Devon border are owned by the Royal Society for Nature Conservation.

The present position is that there are 61,750 acres of woodland and scrub in Cornwall, roughly seven per-cent of the land area of the county; of this 12,350 acres are either plantation or fully managed woods (County Council figures). Assuming one third of the total (say 20,000 acres) to be scrub, we are left with 29,400 acres as a fair estimate for the area of natural unmanaged broad-leaved woodland in the county — total acreage of all woodlands other than scrub is 41,750. Earlier figures of 27,985 acres in 1924 and 34,528 in 1948 (when ninety per-cent of Britain's timber needs had to imported) show that there has been a steady increase in the overall woodland area during the past fifty years.

Commercial forestry in Cornwall, which dates from the early 1920s, is in the hands of the Forestry Commission, the Duchy of Cornwall, Economic Forestry Ltd and a number of private landowners, all of whom have generally operated on broadly similar lines. Initially, commercially valueless oak coppice was cleared and planted with Douglas fir on better sites, and sitka spruce in valley bottoms. Sitka spruce was also planted on moorland and other exposed situations as well as on derelict land where no trees previously existed; such sites, where soils were poor and shallow, were suitable for nothing else. During the second World War, when the war effort and the need to reduce imports were paramount, land was clear-felled and replanted with Scots pine, virtually the only seed then available. By the 1950s, more sophisticated planning could be introduced with planting (including some hardwoods) according to regular annual quotas. Meanwhile foresters have become more conscious of the importance of conservation.

The Duchy of Cornwall Woodlands' programme, mainly in the Ladock, Hessenford and Stoke Climsland areas, has been based on the planting of fifty acres annually, with greater use of American species such as western hemlock, western red cedar and a very small number of coast redwood; these are often planted below oak coppice to avoid the need for unsightly clear-felling. Instead of planting up large stands with a single species, two or three have generally been used. Existing hardwoods have been left on the skyline and along stream-beds while others have been planted on a small but increasing scale: mainly sweet chestnut, beech and the recently introduced,

Woodland Fungi (MB)

Honey fungus   Boletus   Fly agaric   Stinkhorn

fast growing South American beech or raoul. Today forty or fifty per-cent of hardwoods have to be included in any planting scheme if a similar proportion of old trees cannot be left standing. Oak is difficult to grow commerically in Cornwall — as it is subject to 'shake' — except as a coppice crop to produce fuel for wood-burning stoves, and is rarely included; ash, sweet chestnut and beech are preferred. The Duchy also plants American red oak, Norway maple and alder for amenity reasons. Ash can be marketed at all stages of growth, the early trimming being useful for fencing; conifers are in constant demand for pulpwood, building timber and farm use.

Modern conifer rotations are usually 35 years for larch and 50 to 60 years for other species. For hardwoods, other than oak, the earliest felling period is 100-150 years. Oak coppice, managed for fuel on a twelve or fifteen year rotation, seems likely to become more general than in recent decades. These figures are relevant to the wildlife, as both conifers and hardwoods when grown commercially go through a thicket stage — between the sixth and fifteenth years after planting according to species — when low-growing flowers are crowded out. But after the first thinnings have been carried out, the canopy is broken, light enters the plantation and ground vegetation begins to re-appear. Once the rotation is operating properly, the various sections of a modern forest complex should all be at different stages of growth, with excellent diversity for the wildlife.

The forester's aim is to produce a continuous supply of timber of all sizes. To this end the Duchy of Cornwall Woodlands has 2,500 acres of plantation timber operated on the basis of a fifty year rotation. If everything goes according to plan there should be, at any given time, fifty acres one year old, fifty acres two years old and so on up to fifty years. This is ideal for many forms of wildlife, providing conditions suitable for an optimum number of species. Mown rides are essential if brambles, gorse and coppice are to be kept out of the plantation. They allow flowering plants to flourish and act as flight corridors and feeding areas for birds; they are also essential for butterflies such as the brimstone which favour woodland verges. Most fir-cones are produced alongside rides which is also where the crowns of trees grow to their fullest extent. Fungi thrive in plantations and on rides, and are eaten by squirrels as well as by other rodents. The most frequently seen woodland fungi are of the *Boletus* and *Russula* groups; also found are the common puffball, fly agaric which is poisonous, honey fungus and stinkhorn.

A mixed conifer plantation is not attractive to birds during its first few years, when trees are small and well spaced. But when the thicket stage begins, blackbirds and song thrushes appear and warblers become plentiful: willow warbler, garden warbler, blackcap, chiffchaff and whitethroat are the most numerous. As thinning starts, chaffinches and goldcrest are to be seen; a few thrushes and blackbirds remain but some of the warblers move on. Half-way through the rotation many other birds find conditions to their liking: jay, magpie, sparrowhawk, tawny owl, mistle thrush and other species. Wrens and robins are plentiful along plantation edges and where old hardwoods are left standing. Redpolls now nest in conifer plantations and siskins visit them in winter.

Badgers are little disturbed by the replanting of woodlands but foxes prefer the thicket stage. Small mammals tend to remain on the edges of conifer plantations, usually where there are easy access to water and at least a few hardwoods. Fifty years ago there were no deer in Cornwall except in parks. Now red deer are well established in several mixed plantation areas, notably Pencarrow, Herodsfoot, and the Swannacott woods in north-east Cornwall. A few roe have crossed the Tamar and some fallow deer escaped from parks are known to be living wild, but reports of other species must be treated with suspicion. The immediate effect of grubbing out old broad-leaved woods and planting any new species is to reduce markedly, but not necessarily permanently, the number and variety of invertebrates, as many species are associated with dead wood and fallen trees.

In the early years of modern plantation forestry, mistakes were undoubtedly made, both in terms of afforestation and of wildlife conservation. These cannot be rectified at once. A crop has to be taken, which means waiting fifty years or so; also, before growing a mixed forest of the kind now being planted, the right soil has to be established. Foresters have to think in terms of decades or even centuries to achieve their aims. It is likely that the new forests, sympathetically managed with carefully retained or planted hardwood areas, will eventually support an abundance of wildlife at least comparable to the coppiced woodlands of old, though there will inevitably be casualties before this happens. The public tends to be impatient, misinformed and more critical than it need be. The forester asks his critics to share a little of his own patience.

102   Badger in a Cornish Wood. (FC)

PREVIOUS PAGE: Ancient coppice — Nance Wood near Portreath. (NCC) LEFT: Lanhydrock park and woodland; (CGB) RIGHT: Coastal woodland — Valency valley near Boscastle; (CW) and BELOW: pasture and woodland adjoining Boconnoc Park.(NCC) OPPOSITE: Polypody ferns and lichens decorate an ancient tree in Boconnoc Park. (NCC)

LEFT: A woodland ride at Cotehele on the Tamar; (CGB) RIGHT: Two
stages of growth in a conifer plantation where some hardwood still stands,
(CGB) and BELOW: the Wood mouse, or Long Tailed Field Mouse, the
commonest small rodent in Cornish woodlands. (JB & SB)

ABOVE: Spotted Flycatcher — breeds in Cornwall in small numbers; (JB & SB) LEFT: Treecreeper — fairly common in Cornish woodlands, (JB & SB) and RIGHT: Nuthatch — a common resident in the wooded areas in East Cornwall. (JB & SB)

ABOVE: Red Deer in velvet — well established in certain Cornish woodlands: (JB & SB) LEFT: Roe Deer, recently established in East Cornwall in very small numbers, (JB & SB) and RIGHT: Fallow Deer.

Farmlands

Treweege Barton — farmhouse, and pool dug out
in a marshy valley. (FC)

Farming began when man first invaded the wildwood to find land for his primitive crops and
grazing for his animals. Enclosures, in the form of rough stone walls to provide protection from
wind and prevent animals straying, must have followed soon afterwards; indications of this can
still be seen in the pattern of small fields which survives in places, particularly in west Cornwall.
Enclosing land seems always to have been piecemeal, with fields varying in size according to
landscape features and the need for shelter as farms were gradually carved out of the waste. Even
the Enclosure Acts of the 18th century, which changed the face of much of Britain, had little
influence in Cornwall. Again, the typical English arable fields, with regulated use of common
land, were rarely part of Cornish agriculture; neither the 'stitch' cultivation of Forrabury nor the
strips which still surround Kilkhampton were in any way representative.

Cornwall was well settled by Domesday, and most of the farms marked on modern maps were
probably in existence before the end of the 13th century, though the land was not fully cultivated
— waste land, however, was an essential element in the medieval economy, providing fuel, timber
and pasturage. Climatically this period (1100-1300) was mild; more land was taken into
cultivation, and the size of herds and flocks increased. The Black Death (1348) and a deteriorating
climate reduced both human population and pressure on the land so that, in 1540, it was possible
for John Leland to describe Cornwall, the least intensively cultivated county in the south-west, as
'a series of cultivated oases in a large expanse of moor and waste'. The small black native cattle
were still in use — Devon Reds were introduced in the 18th century and Channel Island breeds in
the 19th — while the little moorland sheep produced wool so coarse that it was known as 'Cornish
hair'.

Throughout most of the 18th century Cornish moorlands were largely unenclosed though
extensively grazed. The agricultural emphasis was on pasture, in answer to the growing demands
of local weavers and graziers from beyond the Tamar, but considerable quantities of wheat and
barley were also grown, notably in south-east Cornwall, to supply the Plymouth market. Potatoes
and other root crops were increasingly planted, pig-keeping was becoming more general but, well
into the 19th century, much farming in Cornwall was still at subsistence level. The working animal
was the ox. The traditional system involved a ley-arable sequence with much variation according
to soil conditions and the state of the market, in the amount of land under crops and grass. A
typical rotation, around the year 1800 and for many years thereafter, was two or three corn crops
(wheat, barley and oats) followed by turnips or potatoes (less commonly mangolds, swedes or
kale) then grass, initially sown below cereal, for three to eight years. There were natural meadows
on many farms. The main manures were sea-sand, sea-wrack and rotten fish, all of which had
been used for centuries, while the numerous indigenous clovers no doubt played their part in
maintaining soil fertility.

Today *(County Structure Plan, 1979)* 86% of the land area of Cornwall is in agricultural use, three-quarters of this being under grass or classed as rough grazing; most of the remainder is under cereals. Farms are slightly smaller than in the rest of Britain: 70% under 100 acres and 28% under 20 acres, 75% of dairy farms having less than twenty cows — recent changes in the pattern of farming, particularly in relation to bulk collection of milk, mean that these figures are already somewhat out of date. With so much agricultural land, farming is inevitably one of the major factors in the conservation of wildlife. Misunderstandings and some conflict of interest are unavoidable. Farming is the life-blood of the country. Economically it is one of Cornwall's three main industries — tourism and china clay are the others. These are facts which conservationists should never forget or ignore. There is room for wildlife in the agricultural scene and few farmers would have it otherwise, but the pattern of farming has changed markedly during the last half century with mechanised equipment replacing horses and oxen and human labour, chemical fertilizers taking the place of natural manures, insecticides and herbicides controlling 'pests' — and inevitably other forms of life at the same time.

The right balance is difficult to achieve. At times the conflict of interests is obvious. The farmer may want to drain a marsh where some rare orchid grows. To conserve the orchid's habitat the marsh must be left undrained, and understandably the farmer sees this as depriving him of potentially productive land and increased income. Unfortunately no funds yet exist from which a farmer can be compensated, and his economic problems are difficult enough without this complication. Removal of hedges and reclamation of land covered with gorse and heather are operations which also cause controversy: the farmer needs to create more economic working units, yet these are two of the habitats which naturalists want to preserve. Conservationists argue, in some cases, that the land is too acid for sound agricultural productivity and, long term, that too much tampering with nature is likely to be as damaging to the farmer as to the wildlife. Some conservationists suggest that the answer to many such problems is to bring back horses and switch to biological farming. Certain aspects of biological farming are thoroughly practicable, but the cost of keeping horses is high in terms of both land and cash; even so, there is some revival of interest in the agricultural horse where speed is not a critical factor. Mechanised farming, however has come to stay.

Foxes are accused of killing lambs which they seldom do, most of those found dead being still-born or the victims of badly trained dogs. Stoats and weasels kill more young rabbits than pheasants or partridges; birds of prey and owls do much more good than harm. Yet a few farmers in Cornwall still shoot buzzards, which is illegal, not realising that they are valuable allies in the struggle to control the rabbit population — rabbits were once treated as a useful source of food and income and were even 'warrened' for that purpose. Too many seed-eating birds can certainly be a nuisance but insect-eaters help to control pests in the cornfields. Many insects are harmless, among them the moss carder bee which tends to nest in hay fields. Certain beetles feed on aphids which can be a menace to crops. Most butterflies associated with farmland now depend largely on hedges but, where meadows are still maintained for sweet hay, as in the Kensey and Inny valleys, green veined whites, small pearl-bordered fritillaries and the ubiquitous meadow browns are plentiful. The clouded yellow, a migrant from southern Europe, appears in varying numbers but can be abundant on the south coast in good years.

The harvest mouse is usually described as being a casualty of modern farm equipment, particularly the combine harvester. According to Stephen Harris in *The Secret Life of the Harvest Mouse* (Hamlyn, 1979) there is no evidence to support this view. This unobtrusive little animal was no more widely distributed in the 1800s than it is now. Britain is on the fringe of its range; it is most common in the south-east and east of the country and has never been plentiful in Cornwall where it is now widely but sparsely distributed, mainly in reed-beds, rushes and the long grass of marshland verges. Skylarks might well have suffered from changes in farming practice but have remained plentiful and breed throughout the county. In Cornwall they prefer grass fields and rough pasture to land disturbed by the plough.

The problem animal of recent years has been the badger, normally a thoroughly beneficial species which has been given a degree of protection by law. A county survey made during 1966 indicated more than 1,000 setts, taking the sett as a group of nearby but not necessarily continuous holes used by one family; a figure thought to indicate a population of some 3,000 animals, an estimate which may well have been too low. Three-quarters of the setts were in woodland or scrub, on the coast as well as inland, the rest being on moorland or agricultural land which was visited by badgers from all other habitats; many setts were in or under hedges between different habitats.

In 1971 a dead badger, heavily infected with bovine tuberculosis, was found in an area where cattle were also infected. Infected badgers were then identified in Penwith and in the Morwenstow and Lostwithiel areas. A similar situation existed in other south-western counties, and it seemed to indicate that badgers were acting as a reservoir for a disease lethal to cattle and dangerous to humans. Since then the problem has been to establish the facts and devise a practical policy. It is now accepted that a link exists but exactly how badgers infect cattle is not at all clear. Indeed, it is easier to trace a connection in the opposite direction. Badgers eat beetles found in cow-pats which they habitually search. They can contaminate grassland with their faeces, urine and sputum but, as the tubercle is quickly killed by sunlight, this can hardly be a major route of infection. Badgers can obviously infect each other in a number of ways and probably spread the disease when fighting over territorial boundaries. But this is not the main issue.

After various experiments and a period when badgers were trapped, the Ministry of Agriculture decided early in 1976 to gas setts (using Cymag) to eliminate all badgers in infected areas. This appeared to be the most humane and practical way of dealing with the situation, provided that farmers were not encouraged to take the law into their own hands. The Trust and other conservation bodies reluctantly accepted this situation, hoping that a reduced badger population would be healthier and less susceptible to disease. Nevertheless it is virtually certain that TB, whether in cattle or badgers, cannot be eliminated by gassing badgers. Certain setts, moreover, are inaccessible. Others are of great size. Some individuals are bound to be away when the sett is being gassed. Finding its home sett uninhabitable, a badger will start wandering and searching for a new territory. Fights result, and if a homeless badger happens to be infected, the disease is spread to a new neighbourhood.

There is still much to learn about the behaviour of badgers in relation to TB, the routes and sources of infection in both cattle and badgers as well as between badgers and cattle. Writing in *The Observer* (27 September 1981) Phil Drabble quotes Ministry figures: out of 1,099 outbreaks of TB (by no means all of which were in Cornwall) between 1972 and 1978, 175 were ascribed to cattle imported from Ireland, 331 to badgers, 147 to human sources with 446 shown as 'unknown'. He then asks what happens to cow dung on a farm? Whether or not the cattle are infected it is put into a manure-spreader and spread over the pastures. It would be difficult to devise a better way of infecting not only cattle and badgers but foxes, rats and moles, which are also susceptible. This is just one of the imponderables and may explain at least some of the TB outbreaks of unknown origin. The spreading of cow dung is right and proper. The case against badgers has not been proved. Many healthy animals are being killed for no good reason, and more research is needed.

Many forms of life live largely or even entirely on farmland, and several have become casualties or partial casualties of modern agriculture. Once common flowers of pasture-land and arable are now almost entirely confined to waste places and verges — for example charlock, self-heal, meadow-sweet and poppies with cornflower, never at all common in Cornwall, now virtually extinct. Of particular interest in Cornwall is the heath lobelia, with its attractive purplish-blue flowers, which is now restricted to five or six sites in Britain, one of them on damp pasture in south Cornwall. When the lobelia, which fortunately grows on National Trust land, first came to the notice of the Trust shortly after its formation, there were some 20-30 plants. The site was then accidently ploughed but, far from this being the disaster anticipated, 250 plants appeared the following year — the heath lobelia needs occasional disturbance to survive. There has been some decline since then, but the colony still thrives.

With so much wildlife depending for its survival upon what happens on agricultural land, it is important that conservation and farming interests should learn to understand each other. Those who are actively concerned with conservation of wildlife need to accept that many farmers are equally dedicated. They have created the rural landscape — a dynamic landscape which is always changing — and they are the people who actually live in it. But farming is an industry, and as such must be profitable to survive. A degree of compromise between the farming and wildlife interests is often possible and to help bring this about a Farming and Wildlife Advisory Group (FWAG) came into being. A Cornwall branch was formed in 1977 'to encourage farmers to undertake small conservation schemes; to advise on wildlife habitats and ways of conserving them with minimum disturbance to efficient farming, and to advise farmers who wish to include care of the countryside in their farming plans'. Members visit farms if asked to do so and give advice. The idea is to help, not to instruct, and to see how the needs of modern farming can be reconciled with the conservation of nature. Fortunately it is possible to run a farm productively while retaining some of the best features of its natural history.

The way FWAG functions is best explained by summarising the suggestions made on one Cornish farm, which has the natural advantages of a stream and an old quarry. These were to maintain tree-cover by the stream; to provide a stock-proof fence at a watering area; to plant trees (such as willow, alder and oak) in the quarry, thus creating a valuable habitat without impinging on the working area of the farm — being on the north side of the adjacent field, the trees would not shade productive land but would provide shelter for livestock; to preserve an old lane and avoid close trimming of hedges; to fence off an old beech wood on steep ground to exclude cattle; to leave a strip of unmown grass for hares when fields are cut for hay, as this would also benefit ground-nesting partridges and increase the value of the farm as a source of game. On another farm the advice was to convert a small wet hollow into a pond and encourage biological control of pests by natural predators, birds of prey and small carnivores..

In Cornwall, where so much agricultural land is under grass and farming is largely based upon livestock, the problems are not as acute as in some arable areas, but the planting of winter wheat is increasing, which means fewer bare fields for birds, and economic pressures vary — as do both farmers and conservationists in age, outlook and experience. In one farm of 270 acres in south-west Cornwall, the right sort of balance has been achieved to a remarkable degree. The farmer has improved his land by drainage and hedge removal wherever this has seemed economically sound but, in spite of clearing some 2,000 yards of hedge, his fields still average no more than seven acres. Marginal land which could not be improved economically has been managed for wildlife by digging out pools, forming mud scrapes to attract wading birds, flooding some wet areas and leaving others for rough grazing. To compensate for the cleared hedges, spoil has been dumped on a derelict mining tip where trees have been planted. Trees have also been planted in field corners, around buildings and on rough ground. The remaining hedges have either been left untrimmed or are trimmed lightly every ten years or so — not too long a stretch of hedge in any one year.

The result is a viable, productive agricultural unit on which 35 species of bird nest and over 130 have been seen, including such rarities as red kite, marsh harrier, little bittern, hoopoe and snow bunting — the number of breeding birds would have been greater but for the altitude (550 feet) and the limited area of woodland. Ten species of dragonfly have been noted. Frogs, toads and slow-worms are plentiful, adders fairly common and mammals numerous, including badgers, foxes and the harvest mouse. 'Large farms should be able to set aside a reasonable area for conservation, and I feel it is their duty to do so,' says this farmer, adding that sprays should be used no more than necessary, nest boxes for owls should be fixed to buildings and trees planted as shelter belts.

On another family farm, towards the northern extremity of the county, there are steep slopes as well as partly wooded valleys and stream-beds on both sides of the farm. The main activity is intensive milk production and home-breeding of heifers as replacements for the herd. Both winter and spring barley are grown as regular crops, with hay and beet for feeding cows during the winter. About 100 acres, almost half the farm, is used by cattle throughout the spring and summer with barley as an interim crop. Where a rotation is practicable, it usually takes the form of three years' grass, followed by three years' corn. Herbicides are used with care so as to safeguard hedgerow plants, but the grassland flowers have inevitably been reduced, though pansies, corn spurrey, fumitory and occasionally cudweed still occur in damp places; speedwell, common mouse-ear and chickweed grow in pastures where both pheasants and partridges breed and thrive. Hedges are never trimmed too severely. The result is not only useful wildlife habitats but established routes, along which many species can move from one place to another.

With steep valleys on both sides of the farm, there is a good deal of rough ground. Though some acres have been reclaimed, enough gorse and blackthorn scrub remain to provide habitats for foxes, rabbits, badgers and such small birds as linnets, pipits and stonechats. There are damp patches, a wide uncut hedge and a stretch of willow carr which is used by numerous birds. There is also one steep field which has twice been cleared of gorse during the past ten years. As the gorse is now growing for the third time, the field will be allowed to revert to nature, in which condition it will still provide useful grazing for young cattle. This is a good example of an area which probably cannot be farmed profitably and which is best treated as some sort of natural area or reserve. It is hoped to plant broad-leaved trees on another smaller, steep patch of ground.

This farm is economically viable. It is also an admirable reservoir for the wildlife of the countryside. Most of the commoner mammals are to be found on it including hares, which have always been commoner in east Cornwall than in the west, three species of shrew, two species of bat (pipistrelle and another not identified), harvest mouse, wood mouse and bank vole. There are hedgehogs and moles and, of course, the earthworms upon which they feed; these play an important part in aerating and churning up the soil. Frogs and toads are widespread; palmate newts occupy a small pond. Adders, grass snakes and slow-worms are present; magpies have twice been seen flying away with the last named.

These two larger-than-average Cornish farms are exceptional in that the farmers concerned know exactly what they are doing and why, in terms of wildlife conservation as well as profitable farming. The farms themselves are by no means untypical, though often this is accident rather than design. Details are particularly important in terms of wildlife survival: the timing of hedge-trimming operations, for example. There are also farms where — at least in the view of conservationists — the land is plundered in the interest of short term profits. The farmer does not necessarily see it that way. He sees no reason why he should be expected to take from his farm less income than it would otherwise yield in order to pay indirectly for wildlife conservation, which he considers to be of benefit only to others. This may be a wrong-headed view, but it is also one of the biggest single causes of misunderstanding. That a reasonable part of the wildlife of the countryside should be preserved is a matter of universal interest.

Dabchicks find a home on a farm pool at Treweege Barton.
(FC)

ABOVE: Sheep grazing on the Predannack Downs, the Lizard, (NCC) and LEFT: ponies and sheep on Bodmin Moor grassland. (JB) RIGHT: Treweege Barton, Stithians — uneconomic farm land used for conservation; (JB & SB) OPPOSITE ABOVE: arable encroachment upon the cliff lands — cultivation right to the edge of the cliffs. (JB) CENTRE LEFT: A female Mole Cricket — a rare insect which may occur on damp meadows in Cornwall; (CGB) RIGHT: the flightless Violet Ground beetle is generally common; (CGB) BELOW LEFT: the Small Copper Butterfly associates with Docks and Sorrels, (CGB) and RIGHT: the Musk Beetle, less common than formerly, occurs in damp places. (CGB)

115

ABOVE: Badger, the problem animal — is it really responsible for the spread of tuberculosis? (JB & SB) LEFT: The Common Frog is scarcer than it used to be; (CGB) CENTRE: the Common Toad is frequent in pools and damp places, (CGB) and BELOW: a mole, seen eating earthworms. (CGB)

Hedges and Verges

Well grown Cornish hedge — magpie in flight. (FC)

In Cornwall the term 'hedge' usually means a dry stone wall or a double wall with earth in between and turf on top. The stone used varies with the locality: granite on Bodmin Moor and in Penwith, serpentine on the Lizard, pillow lava around the fields of Pentire Head, limestone in parts of south-east Cornwall and elsewhere, usually slaty rock or sandstone. Ferns and flowering plants soon establish themselves on such hedges which, except in exposed situations, eventually become so well covered with turf that they look like banks — lichens decorate bare exposed hedges along with wall pennywort and other members of the stonecrop family. Woody plants, led by gorse and brambles, soon appear but you cannot date hedges in Cornwall by their woody plants as in many other counties, though diversity increases with age. Trees — mostly oak, ash, hawthorn, hazel and holly — have evolved in many hedges, and tradition used to assure that they were left and replaced where necessary; cutting down holly, or 'Christmas', was believed to bring death to a farm horse. Today, with mechanical cutting, hedgerow timber trees are rarely left to reach maturity, and the loss to wildlife is considerable. In Cornwall, this may well be more serious than the loss of hedges.

There are roadside hedges, hedges beside ancient sunken lanes and hedges between fields, farms, manors and even parishes; there are also hedges placed to protect or shelter farm animals. Some are ancient and may date from the Bronze Age. Parish boundary hedges were established more than a thousand years ago. The pattern has developed gradually, almost as a natural process. Even today, hedges are being made alongside new roads and realignments, and this is much to the credit of the County Council and the men who still take pride in building beautiful old stone walls — not for Cornwall the mean fences which bound the new roads of so much of Britain.

Road verges are not the same as hedges, though the two overlap and their conservation is part of the same problem. On a national scale it is reckoned (by a panel of local authority ecologists) that 700 plant species grow on road verges and hedges; of the 300 rarest British plants, 27 occur on hedges and for eight this is their main habitat. All British reptiles, 40% of mammal species, 20% of birds, 42% of butterflies and 47% of bumble-bees breed in verges and hedgerows. Subject to minor variations these figures may be accepted for Cornwall. The sheer volume of wildlife which can be affected by spraying and the use of flail mowers is frightening in its magnitude.

If for reasons of road safety cutting cannot be avoided, no harm is done if the yard or so adjoining the road is cut regularly, provided the middle section is cut only once annually in the autumn, and the back section, which in Cornwall includes the hedge itself, is cut in the autumn only after an interval of some years. If trees and shrubs are allowed to take over open verges, maintenance is largely obviated as, for example, on the new St Columb by-pass, where there is a

Flowers of a Cornish hedgerow (MB)

Pyrenean cranesbill

Bastard balm

Sweet violet

Goldilocks

Marsh orchid

Columbine

Wild cherry

Himalayan balsam

118

Red campion

Cow parsley

Herb robert

Stitchwort

Buttercup

Primrose

Violet

Celandine

(MB)

Dandelion

Wood sorrel

119

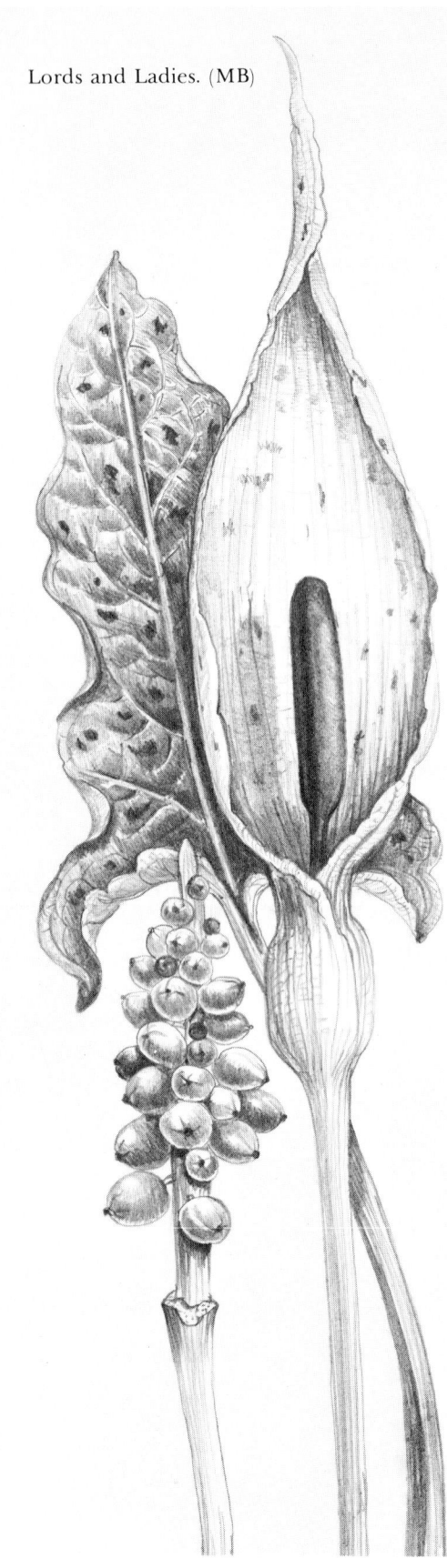

Lords and Ladies. (MB)

fine display of gorse, or where broom flourishes. Work ought to be planned with the natural cycle in mind. The main growth period for hedgerow plants and shrubs is from April to early August; they fruit between July and October, providing food for many mammals and fruit-eating birds at least until November. The nesting period for most hedgerow birds lasts from early April until June. Insects, upon which many birds depend, are most numerous between May and August. Knowledge of what is growing or living in any given hedge enables the most vulnerable periods to be avoided.

Cornwall is extremely fortunate in its county surveyors who have an excellent understanding of these problems. In the late 1960s, the Trust identified sections of road verges where rare or unusual plants were growing and where there was a particularly fine display of common species. These were marked 'NR' (Nature Reserve); the Trust gave advice, and they were treated accordingly. Among the plants were Pyrenean cranesbill on a lane near Marazion, Cornish heath (away from its home on the Lizard) on Connor Downs in Penwith, Dorset heath and marsh orchid beside the A30, bastard balm near Ruan Lanihorne, broad helleborine near Callington and wood goldilocks on the road from Bude to Morwenstow. Sweet violet, Cornish bladderseed, western fumitory and the large-flowered mullein were also protected — the early purple orchid is still quite plentiful in Cornish hedges. One of the best collections of interesting, but not rare, plants was on a lane near Kilkhampton where Himalayan balsam, columbine, bridewort, soapwort, guelder rose and wild cherry still grow in some profusion. A number of ferns, including black spleenwort, polypody and golden male fern also occur on many hedges and verges in Cornwall.

In 1975, fifty-seven Women's Institutes took part in an eight-month survey — organised as a competition — of ten yard stretches of hedges and road verges all over Cornwall. Taking yard-square grids, divided into six-inch squares, the institutes filled in forms covering the remarkable total of 70,000 squares. Red campion was the most numerous and widespread of the flowers recorded. Buttercups were plentiful on the verges in May but relatively few primroses were noted, partly because of the late start of the survey and possibly also because 1975 was a dry summer — primroses are still plentiful in many Cornish hedges. Other plants noted frequently in both hedges and verges were celandine, stitchwort, herb robert, bluebell and sorrels with violets on hedges and hogweed, dandelion and cow parsley on the verges. In all, 168 species were recorded. Few institutes reported cutting before June; almost half the hedges, as opposed to verges, were left uncut. But July trimming often left the roadside bare of flowers for the rest of the year. If the survey could be repeated, long-term trends, in both numbers and variety, would become evident.

The animal life of Cornish hedges and road verges is abundant. Bank vole, wood mouse and both common and pigmy shrew habitually make their homes in hedges, as occasionally does the house mouse. Weasels make use of holes between the stones, and rabbits burrow into gaps or below the stone-work. Badgers frequently dig out their setts below Cornish hedges. Hedgehogs find shelter and sometimes make their nests at the base of Cornish hedges, and are worth looking for in the thick growth of the wild arum, lords and ladies, a common hedgerow plant often found in association with the stinking iris, which has bright orange seeds in the autumn. Snakes and lizards also make good use of hedges. Perhaps the most typical of the many hedgerow birds are robin, chaffinch, dunnock and wren, the latter frequently nesting between the stones. These and many other birds feed their young on insect larvae and are seriously endangered by the use of insecticides. Where trees are allowed to grow in hedges, woodpeckers, nuthatch and owls find nesting sites and food supply close together.

Many insect species are largely or totally dependant upon stinging nettles, which commonly grow around old hedges. Several of our most admired butterflies are associated with nettles: peacock, red admiral, painted lady, comma and the small tortoiseshell, which is a favourite food of the dunnock. The common blue, which so often gives rise to false reports of the large blue being seen again on the wing, is frequently observed in summer, flying along hedges in fields or open country. The hedge-brown or gate-keeper, a reluctant flier, rarely moves far from the hedge, where it spent the larval stage of its life; wall browns sun themselves on stone hedges, and the ringlet takes nectar from bramble flowers. All these butterflies are still common in Cornwall. The marbled white has suffered greatly, in Britain as a whole, from hedge cutting and spraying and has become distinctly rare. It is still well-distributed along Cornwall's eastern border but has never been common in the western part of the county. The caterpillars of many moths eat the leaves of hawthorn and other hedgerow plants but most are nocturnal and seldom seen. The fine oak eggar feeds along hedges by day and rests in their shelter. The large emerald, a night flier whose caterpillar is unusually well camouflaged, rests on hedges by day and is worth looking for.

Many hundreds of other invertebrates live or spend part of their lives in hedges. Many people think there are altogether too many of them for man's convenience, but they can be of much interest in themselves and play an essential part in many of nature's food chains. For instance, the kestrel commonly hovers and hunts over road verges where it kills small birds or mammals which live in hedges feeding upon insects from the same habitat; these in turn feed upon hedgerow plants. If the kestrel is killed by a car the carcase — after being partly eaten by ants, various beetles, fly larvae and other small creatures — will eventually be absorbed into the soil to provide nutrients for plants. The ubiquitous slugs and snails — the garden, dark-lipped and white-lipped snails are the commonest species in Cornish hedges — are eaten by thrushes, part of another hedgerow cycle.

Small tortoiseshell

Peacock

Red Admiral

Comma

Butterflies on stinging nettle (MB)

Hedges and verges are favourite habitats for several species of bumble-bee, which visit flowers for their nectar and pollen and play an important part in the fertilisation of many plants. The small earth bumble-bee is common throughout Cornwall, often nesting in holes in hedges abandoned by bank voles. Another widespread species is the early nesting bumble-bee, which is one of the first to appear in spring, when it feeds on blackthorn and other early flowers. At times it occupies disused birds' nests and is itself parasitised by Barbut's cuckoo-bee. Honeysuckle which decorates many Cornish hedges is regularly visited by the small garden and short haired bumble-bees; the latter is nowhere common but has a wide range in the county.

The common green grasshopper is frequent whenever there is lush vegetation on roadside verges. The larger and darker common field grasshopper is also likely to be seen. Dark bush-crickets, which are almost wingless, crawl along hedges and chirp in chorus from their hiding places among brambles and nettles, particularly in the autumn. Among a host of other creatures, there is a large number of spiders many of which live in hedges and the associated vegetation. One of the commonest is *Segestria senoculata,* which builds its silken and tubular retreat in crevices in walls and rocks. Another much less common member of the same genus, *S. bavarica,* builds in cracks in rocks near Tintagel. Garden spiders also build in hedges, where their webs are a familiar sight late in the summer. The long-legged harvestmen, such as *Phalangium opilio,* one of the commonest, clamber around with great agility at the base of hedges. Beetles are numerous and include among their number the charming two-spot and seven-spot ladybirds; both frequent hedges as well as gardens, where they should always be made welcome for the aphids they consume.

There are many hedges in Cornwall worthy of individual description, from those in exposed coastal situations — with their tufts of thrift, clumps of stonecrop, ferns and lichens and, perhaps, a colony of digger wasps — to those beside ancient sunken lanes. Two of the finest flank an old lane which leads from Treworra farm to Davidstow Moor. The high-banked hedges are crowned by gnarled and wind-blown trees: mostly oak, elder, hawthorn and beech — some with signs of past coppicing. The flora shows the influence of both shade and altitude, which is marginally below 1,000 feet. Ferns are abundant and include both broad buckler fern and polypody growing on trees. There are more than thirty mosses and liverworts, six of which are supported by trees; others grow on boulders. The beard lichen is also abundant, and there are plenty of fungi and galls. Among the flowering plants are wood sorrel, wood avens, violets, wood vetch, celandine, wild arum, gorse, foxglove, red campion, ground ivy, pennywort, bog stitchwort, marsh thistle and some *Juncus* rushes. There is a rookery at one end and, at the other, an open moor much favoured by flocks of lapwing and golden plover. There is a colony of rabbits, a badger path, a fox's earth, snails, slugs, butterflies and many other animals. The Trust has a sanctuary agreement with the owner, so that this beautiful lane may have the protection it deserves. The hedges of Cornwall are well worth preserving and guarding as a precious asset and part of the county's heritage.

Stoat at the bottom of a Cornish hedge. (FC)

LEFT: A newly built Cornish hedge, (CGB) RIGHT: a single oak left standing in a typical hedge, (CGB) and BELOW: a granite hedge near the Great Pool of Tresco on the Isles of Scilly. (NCC)

LEFT: The ancient sunken lane at Treworra Farm on Bodmin Moor, (JB) RIGHT: Wall Pennywort on an old hedge at Gorran Haven. (CGB) BELOW: trees have taken over the remains of an ancient hedge in Boconnoc Park. (NCC) OPPOSITE LEFT: An uncut Cornish hedge near Penryn; (CGB) RIGHT: Dark Bush Cricket; (CGB) CENTRE: the Scorpion Fly haunts sunny hedges, particularly where there are blackberries or nettles, (CGB) and BELOW: young queen wasps hibernating in the nest where they were raised. (CGB)

ABOVE: Bank Vole, the commonest small mammal of Cornish hedges, (JB & SB) and BELOW: hedgehogs often seek shelter in well grown hedges. (JB & SB)

Heaps and Holes

Soay sheep on a china clay slope. (FC)

There are 11,140 acres of derelict land in Cornwall of which 2,175 acres are considered by the County Council to be so unsightly as to require reclamation. Most of this is the result of past metalliferous mining and the extraction of china clay. Present clay operations are largely concentrated into an area of 17,000 acres around St Austell but, of this, only some 7,000 acres are occupied by active tips, pits, dams, works buildings and so on. There are also china clay workings on Bodmin Moor and smaller workings elsewhere, bringing the latter figure up to an estimated 9,000 acres as actually in current use by English China Clays Ltd. The areas dominated by the industry and to some extent threatened by future development are much greater than this. The problems of reclamation and conservation are formidable.

China clay is not a particularly uncommon mineral. Large deposits are found in continental Europe and North America. It also occurs in China, where it may have been worked as early as 1000 BC and was definitely worked between the seventh and ninth centuries AD in the Kaoling hills in the Kiangsi province of south-east China — the origin of the word 'kaolin'. In Britain it occurs only in Cornwall and in Devon at Lee Moor on Dartmoor. Millions of years ago, pressures created by disturbances in the earth's interior forced hot gasses and fluids through the granite strata. This caused partial decomposition of the granite and changed the felspar (a component of granite) into kaolin or china clay. These 'zones of alteration' have a funnel-like shape and extend down to considerable depths. Other constituents of the granite (mainly quartz, mica and tourmaline) did not change; with the overburden, mainly of rock and peat, they form the waste when the clay is mined. This is the essence of the conservation problem: for every ton of china clay there are seven tons of waste, and if this was to be dumped back into the excavation pits, it would bury the deeper levels of clay.

The Chinese method of using kaolin to make high quality porcelain was kept secret until 1712. Then the search began for kaolin in Europe. Deposits in Britain were first found by a Plymouth apothecary, William Cookworthy, in 1746 at Tregonning Hill near Helston. Two years later, much larger deposits of better quality clay were found near St Austell at St Stephen-in-Brannel where it is still being mined. This was the real beginning of the industry as we know it. At first the clay-pits were worked by established potters from other parts of Britain (Wedgwood and Spode among them) but, after about a hundred years, their leases were relinquished and the pits taken over by numerous Cornish families. Mergers followed. In 1919 the largest producers came together to form English China Clays Ltd. Today approximately 18% of the world's china clay comes from Cornwall and Dartmoor. It has manifold industrial uses including ceramics, tile-making, paints, plastics and paper — after wood-pulp it is the most widely used material in the production of paper. Three-quarters of the output is exported.

Kaolin is extracted from the deposits by high-pressure jets of water which break up the material. The waste (quartz-sand, undecayed granite and overburden) is carried away from the pits by truck or conveyor to produce the huge conical or barrow-like heaps which are such a familiar feature of Cornwall's landscape. The mica and kaolin are pumped into settling tanks where the kaolin is separated out, leaving micaceous residues in the form of a slurry to be dumped into excavations known locally as 'lagoons'. When these are full, vegetation can be established without difficulty, as the material is not toxic and retains well any nutrients applied. Most full lagoons have been successfully seeded for use as pasture, as at Hawk's Tor on Bodmin Moor, but they sometimes develop a tendency to become waterlogged. One such lagoon has been colonised by willows, alders and other wetland plants. Thus lagoons are not a long-term conservation problem. Before 1973 most of these residues were discharged into local rivers, mainly the St Austell (or White) river and the head-waters of the Fal. Though the material was in no way toxic it created considerable pollution in the sea as well as in the rivers themselves.

The waste which has been building up in heaps for over 200 years has been used for making concrete blocks and pre-cast cement products as well as in road construction. But transport costs make it uneconomic to send it further afield than Devon, so that only a fraction can be utilized. Some of the oldest tips have been recolonised by natural vegetation (ling heather, bell heather, gorse, bilberry, broom, brambles, foxglove, sorrels and grasses) but this takes many decades to develop. It is, however, similar to the normal flora still growing on undamaged heathland in the china clay areas — Japanese knotweed is the principal weed of waste places. But modern extraction techniques produce waste of almost pure quartz-sand, which is not only deficient in nutrients, but also extremely porous, so that nothing grows naturally. Even so, with the help of lime fertilizers, a grass-legume sward (sometimes with added clovers and bird's foot trefoil) can be established, without the addition of top-soil which just isn't available in sufficient quantities even if it could be made to stick.

Some tips are unstable and liable to erode. Others develop a hard crust, which increases the limitations on plant growth. All are so steep that conventional agricultural machinery cannot be used for seeding operations. Hydraulic seeding has therefore been introduced by ECC's landscaping section. A slurry of seed, fertilizer and lime, followed by a mulch of peat and chopped straw, has been sprayed on to the tips, though not always with success, owing to the almost total absence of nitrogen in the base material. However, satisfactory swards, mainly fescue and bent grasses, have been established by this method on a few south-facing slopes, as at Bugle, and these are being grazed by hardy Soay sheep, specially introduced to Cornwall for that purpose.

Reclamation of these tips is fraught with difficulties. The old conical tips could not be landscaped and seeded while still used for dumping; the height (over 300 feet) made the upper levels difficult to deal with. The more modern tips are lower (80 feet) and are being constructed in the form of elongated 'benches', which allow vegetation to be established behind the advancing face of an active tip. There remains, however, the essential ecological problem of converting material lacking organic nitrogen and other nutrients into a 'productive and self-reliant eco-system'; this is being worked on by research teams financed by ECC and directed by the botany department of Liverpool University. Meanwhile, and in addition to direct operations on the tips, the company has organised a large tree nursery at Carthew, and some 20,000 trees (mostly hardwoods) have already been planted on sand slopes. More than thirty belts of some 3,000 trees each have also been established as screens to hide unsightly working areas from the public view and to make the local environment more attractive. The Penhedra tip has been surrounded by a twelve acre forest, and the sites of certain new excavations have been effectively screened before work has actually started; new harbour installations at Fowey have been accompanied by a planting scheme involving beech, oak and ash. This is only the beginning of a most laudable effort to deal with one of the largest, most obstinate and complex reclamation problems in Britain.

The position is not static, however. The excavation of china clay will continue, and this means more tips and more lagoons. This must inevitably cause concern to naturalists as must the possibility of expansion into new areas and the extension of work already started, particularly in such situations as Stannon Downs just below the western flank of Rough Tor, one of the most spectacular parts of Bodmin Moor. In the main Hensbarrow area, the probability is that in future several tips will be amalgamated to create larger and higher units, properly landscaped to present a more natural outline, with wooded lower slopes and vegetated upper slopes suitable for grazing. The intention is to create a totally new upland landscape, not to degrade the old. It is a thought-provoking prospect.

The pressures that turned felspar into kaolin millions of years ago also filled cracks and fissures at the margins of the granite bosses with mineral-bearing solutions. These crystallised and eventually formed the lodes of tin and copper which have been extensively mined around the fringes of Cornwall's granite moorlands; and lead, silver, zinc, tungsten and iron, which have also been worked, but on a much smaller scale. The china clay interest dominated Hensbarrow moors above St Austell but Bodmin Moor, West Penwith and particularly Carmenellis have given rise to numerous mining undertakings. Surface tin (*ie* tin oxide, or cassiterite, lying below three or four feet of peat and gravel), exposed by erosion in moorland valleys, was known to the inhabitants of Cornwall from an early date. It was smelted initially with peat and later with wood and remained the only source of this metal until the 16th century. Primitive mining, as opposed to streaming in wet valleys, must have started where lodes were exposed on cliff-faces or steep hill-sides. The method was first to 'trench' into the lodes and later to tunnel out adits (horizontal approaches) in a way that allowed the ore to be extracted from below and simplified drainage. True underground mining, below adit level, could not be developed until efficient drainage became possible with the invention of steam pumping engines in the early 18th century. At this period, too, 'blowing houses' were introduced where the ore was heated with charcoal and 'blown' by bellows worked by water-wheels. The remains of the buildings which housed these 'Cornish pumps' still form a conspicuous feature of the mining landscape. The demand for charcoal and the need for timbering in the mines made heavy demands upon Cornwall's limited woodlands which must have been seriously depleted in areas where coppicing was not adopted. Historians (Thomas Tonkin, 1678-1742) have sometimes suggested that the country between Carn Brea and the sea was heavily wooded at the beginning of the 17th century. But this is unlikely except in the valleys. This area is not now wooded, and there is no evidence of woodland at the time of Domesday. Moreover, coal was being imported before 1700; the demand for charcoal was diminishing and smelters were moving to the coast.

Arsenic was a troublesome impurity in many mines and had to be removed by roasting. This produced arsenic oxide, which was deposited as a white powder and used at one time for spraying the American cotton crop. The smoke produced by roasting was led away in flues and issued from tall stacks, thereby seriously polluting the surrounding countryside — in some areas the arsenic pollution still remains. In the few mines now operating, arsenic is 'dumped' — presumably in safety.

Another development was tin-streaming, quite a different process from that used in ancient times. Late in the last century it became profitable for small operators to treat water, red with oxide from the mines, and extract cassiterite crystals which had escaped the separators. Streaming was mainly but not entirely concentrated upon Red River, between Carn Brea and the sea at Godrevy, and a smaller stream between Wheal Bassett and Portreath. Red River has carried mine residues for centuries. The water is still sterile, but the flat-bottomed valley is partly wooded (birch, willows and alder) and partly marshy with a flora largely dominated by horsetail. The subsidiary channels, cut for tin-streaming but abandoned long ago, now support a plentiful aquatic fauna; dragonflies and butterflies are numerous, and Daubenton's bat appears to find conditions to its liking. But where the valley has been recently worked over for possible tin residues, it seems likely to remain derelict for years.

During the eighteenth century copper mining grew in importance and flourished until about 1860. New mines were opened and existing mines deepened. The centre of this development was around Camborne, Redruth and Gwennap, with other important mines in the St Blazey and St Clear areas. The Levant mine at St Just produced appreciable quantities of ore as did others in the Tamar valley, notably the Devon Great Consols group near Tavistock and Calstock. Copper ore was dressed by hand, the waste rock being dumped in huge mounds or 'burrows', which remain a feature of old mining sites. Unlike tin, copper was seldom smelted locally but was exported in the form of ore. The only serious venture into smelting was at Copperhouse in Hayle, though this was uneconomic and discontinued after a few years. But the works and development of the port which exported its products involved opening up the Copperhouse estuary, which has become an important area for wildfowl and waders on passage. All that remains of this venture are the blocks of black scoria, a by-product of copper smelting, which wall the sides of the estuary, bridge the salt-marsh and enclose the churchyard of St Filius at Phillack.

Ivy-leaved Toadflax on an old ruin. (MB)

Tin is still mined at St Just, Redruth and Bissoe. Interest is being shown in opening up certain old abandoned mines as at Redmoor mine near Kit Hill; a few old dumps are being excavated and moorland areas prospected for alluvial deposits. Dredging for offshore accumulations of ore, washed down to the sea in streams, is now taking place off St Agnes Head and may be attempted in St Ives Bay. The extent of future environmental damage is difficult to estimate. Redmoor shows signs of becoming a major tin producer. Prospects elsewhere are still uncertain.

Waste land surrounds the old mining sites: on Carn Brea, at Gwennap where United Mines operated, and in Bissoe Valley at Twelveheads. There are plundered lands and derelict buildings along the coast between Pendeen Watch and St Just, at St Agnes Head and many other places. It is the same in east Cornwall — along the southern fringe of Bodmin Moor, on Caradon Hill and in the middle Tamar mining region. Newlyn Downs, between Mitchell and Perranporth, is a mass of old 'burrows', heavily contaminated by the Wheal Rose complex of lead and silver mines. Many valleys, which once carried water and iron-oxide 'slime' from the mines, still show signs of the deposits. The old engine houses and other mine buildings interest industrial archaeologists and are not unattractive, particularly when they stand upon the clifftops as at Botallack, St Agnes and Trewavas Head. Ferns, wall pennywort and ivy-leaved toadflax grow in crevices between the blocks of stone. Jackdaws, martins and other birds nest in them; bats, believed to be pipistrelles, use them as roosts — greater horseshoe bats once hibernated in mine-shafts in the St Agnes and Camborne areas and may still do so. The rare six-eyed spider *Segestria bavarica* may sometimes be found in cracks in the mortar of these old buildings.

The rough ground surrounding a few of the oldest mines, abandoned before the copper boom, is almost completely covered with vegetation, the dominant plants being ling heather, gorse and hawthorn. The copper areas suffer from an excessive concentration of sulphides and are still relatively sterile after more than a century. Where they have been colonised, ling heather and gorse are again the dominant species but there is much bare ground. The old tin mines are mostly surrounded by a poor heath flora except where 'slimes' with a high iron content have formed an impervious layer of clay which supports little beyond grasses and sedges. Where 'burrows' have been levelled for re-processing, so that any remaining tin residues could be extracted, the site is quickly recolonised by yellow-flowering coltsfoot, thistles and the usual waste ground flowers, as well as heather and gorse. On the whole the fauna is sparse but these areas are rugged and undisturbed. Various mammals find that they provide safe and convenient refuges. Invertebrates come to the plants with which they normally associate. The orb web spider and the rare *Araneus adiantus* both occur in heather and gorse in these surroundings while the mesh-webbed spider is often found under loose stones.

Though man's influence is felt throughout our modern environment, in Cornwall the most dramatic impact is in the areas worked over for the extraction of china clay and metals. The impact of tourism, however, is more insidious and in the long run could prove more damaging. There are no other major industrial undertakings and no large conurbations. The urban sprawl is not in itself a serious threat, except for the demand for land to accommodate visitors in chalet camps and standing caravan parks, and for retirement homes by the elderly.

Wildlife exists in towns everywhere, to such an extent that built-up areas are coming to be accepted as another wildlife habitat, albeit rather an inferior one. Wild flowers and animals are liable to appear in considerable abundance in such places as abandoned railway lines and old stations, and not only in parks and gardens. In Cornwall, a majority of towns are on the coast and seem to have been absorbed into their surroundings. The wildlife of the coast and shore is part of the background. Sand-dunes adjoin built-up areas; rivers and estuaries pass through them. The Copperhouse estuary in Hayle is one example. St John's Lake (part of the Tamar estuary), Truro river, Fowey estuary, Par sands and the Bude canal are others. Badgers have setts in more than one Cornish town. Foxes move about the streets at night, and herring gulls nest on roofs in St Ives. Many species show a surprising capacity to adapt, and some are primarily associated with human dwellings and buildings: certain bats, barn owls and the daddy-long-legs spider for example.

Delabole slate quarry. (FC)

ABOVE: China clay pit and workings near St Austell, (CGB) and BELOW:
China clay workings — a scene of utter desolation. (CW)

ABOVE: Old china clay spoil heaps, partly overgrown, and a pool, (CGB) and BELOW: Tree lupins growing on old china clay spoil heaps near St Austell. (CGB)

ABOVE: Old mine buildings on the cliff tops — Trewavas Head, (CW) and
BELOW: Barn Owl with Common Shrew — this owl favours old buildings,
and in Cornwall often hunts by day. (Donald Smith)

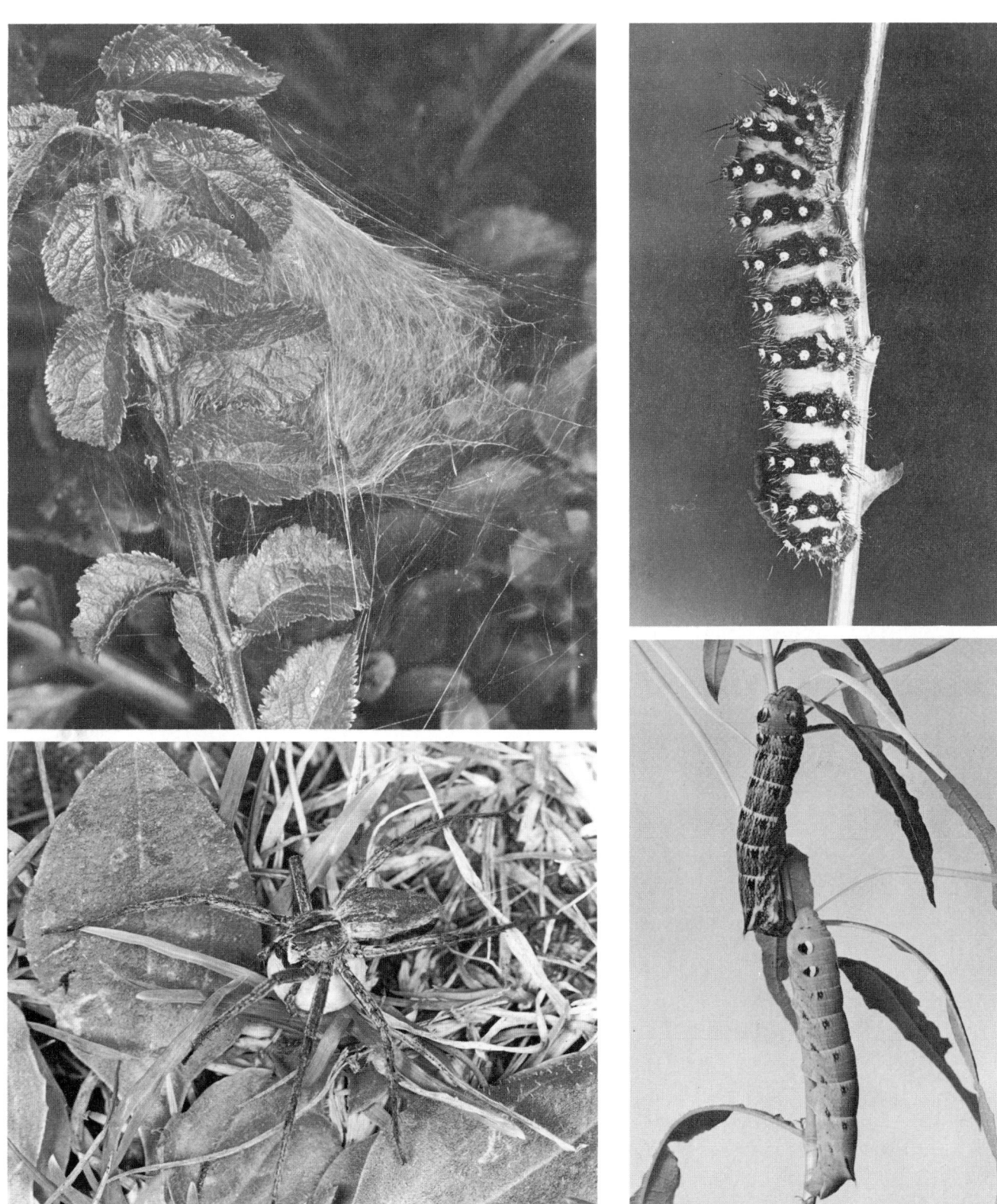

ABOVE LEFT: Web of the spider *Pisaura mirabilis,* often seen on heather, (CGB) and BELOW: *Pisaura mirabilis* spider with egg cocoon; (CGB) ABOVE RIGHT: Caterpillar of Emperor Moth, dark form — associated with heathers, (CGB) and BELOW: both dark and light forms of the Elephant Hawk Moth caterpillars — not uncommon on derelict land. (CGB)

ABOVE: Trethergy Pool, once a china clay pit, now a trout fishery surrounded by natural vegetation, (CGB) and BELOW: old mine buildings near Minions on the edge of Bodmin Moor. (CGB)

Between the Tides:
Life of the Seashore

Rock pool on a rocky shore. (FC)

Around Cornwall's immensely long coast-line there are headlands where the cliffs drop straight into the sea. Elsewhere the cliffs are awash when the tide is high, but separated from the sea at low tide by a wide belt of shore. Cliffs are absent altogether in sand-dune areas, and in the mouths of rivers or estuaries where the strand-line lies along the banks. The shores themselves result from the varied geology of the coast and may be rocky, sandy, boulder-strewn or stone or pebble beaches; almost always there are rock-ribs, pinnacles, platforms, pools, caves, cracks and crevices. Every species of seashore life has its habitat requirements, and geological variety ensures habitat variety. In many quite small bays around the coast, examples of several different seashore habitats are to be found close together. Among areas where this variety can be seen are Bude, Trevone and Harlyn Bays, Porthcothan, Crantock Beach near Newquay and the coast between Gorran Haven and Dodman Point.

Beaches differ markedly in the degree to which they are exposed to the sea, the wind and the weather. Exposure is greatest along the granite shore of the Penwith peninsula, and where the high cliffs and headlands of the north coast protrude into the Atlantic. Even the most open stretches of the south coast, where there are so many bays and inlets, are sheltered by comparison. On the Isles of Scilly exposure is generally great, though the inhabited islands are often protected by outlying rocks and stacks, and eastern beaches face away from the weather.

The amount and type of shelter offered to plant and animal life determine the variety of species likely to thrive on any particular beach. Seashore life has to contend with being first submerged and then exposed to the air — including extremes of rain, sun and wind — for varying lengths of time. Species which tolerate longer periods of exposure live towards the high-tide levels; others which cannot stand being uncovered for more than a short time live where the tide leaves them exposed only when it is right out. The ability to tolerate these changing conditions, brought about by the movement of the tides, varies with the species. For this reason the distribution of species on the shore tends to occur in bands between the levels of the highest and lowest tides. Competition with other species, exposure and the availability of shelter are the main factors determining which species actually survive.

Above the level of the highest tides there is a 'spray zone' on most rocky shores where the orange lichens are conspicuous; closer to the high-tide line another lichen, black *Verrucaria maura*, can easily be mistaken for dried oil. The seaweed living nearest to the high-tide line is channelled wrack, normally yellow but black and apparently lifeless when dry. A little lower down, roughly at the half-tide level, there is bladder wrack with air-bladders; many Cornish beaches are so savagely battered by the waves that the air-bladders are much reduced and sometimes not apparent at all. At this level, on more sheltered beaches, there is a band of knotted wrack; and at low-tide levels there is saw-toothed wrack, thongweed and the tangle or oarweeds which extend into the shallows. There is also a giant annual seaweed known as furbelows which is abundant locally — it is worth looking for the small hairy crab in its knobbly 'holdfasts'. (The above are usually the dominant seaweeds on Cornish beaches, but they are by no means the only ones.)

Channelled Wrack   Bladder Wrack   Saw-toothed Wrack

Knotted Wrack

Common seaweeds (MB)   Thongweed

The different seaweeds may harbour a different range of animals, which makes it possible to look over a beach quickly at low tide and see which animals are likely to be present in association with the weeds. Sloping beaches in a reasonably sheltered part of the coast are most easily viewed in this way, Feock in the Carrick Roads being a particularly good example. A close look at any patch of seaweed soon reveals the animals living on or under it; sponges, worms and molluscs are usually the most obvious. Among the drapery of weeds there may well be rock-pools, with their populations of small fish such as gobies and blennies, as well as anemones, starfish and so on. A rocky shore at low tide, where such pools of sea-water are left behind in fissures and crevices by the receding sea, is much more interesting than a bare sandy beach; on the latter there is little animal life and this is mostly hidden away in the sand.

Cornwall's south-westerly situation in Britain and the way in which it protrudes into the Atlantic are, with the varying degrees of exposure, the dominant influences on the seashore life. The county's shore-line is greatly influenced by the warm water of the North Altantic Drift, an extension of the Gulf Stream. Plants and animals of the warmer south meet those from the cooler northern waters. Some northern species reach their southern limit in Cornwall: the seaweed 'dabberlocks', for example, with its supposedly edible midrib, is common at low tide on many north coast shores. The beautiful iridescent *Cystoseira tamariscifolia*, with a mainly southern distribution, occurs in rock-pools chiefly along the south coast. Seashore life in Cornwall is particularly interesting because of this meeting of northern and southern species with a certain amount of overlap.

There are numerous animals with a south-westerly distribution, two of the most common being the thick topshell — particularly on Scilly where it is regularly eaten — and the snakelocks anemone which is present in many shallow rock-pools. Of the three large limpets found in Cornwall, only the true common limpet occurs throughout Britain. The other two are the dark-footed limpet, which prefers rocks battered by the surf, and the china limpet which occurs in rock-pools in the lower and middle shores; the vernacular name refers to its interior which looks like porcelain. The large perforate barnacle is common in crevices on the lower shore. Among smaller barnacles which completely cover some rock surfaces, the northern *Balanus balanoides* is unevenly distributed in Cornwall and may be becoming scarcer. The southern barnacles are abundant in Cornwall, the former on rocks beaten by the waves and the latter in more sheltered situations higher on the shore. Certain cuttle-fish and octopods are among other southern species which sometimes appear, but the one truly Cornish marine animal is the Celtic sea-slug.

A more direct result of Cornwall's situation and of the ocean currents is the frequency with which transatlantic species find their way to Cornish shores, especially when westerly winds are strong and persistent — such animals are in danger once they reach the colder waters of the North Atlantic. 'By-the-wind-sailors' sometimes arrive in millions as in June 1981 when these jellyfish-like creatures spread round the coasts of Cornwall, Devon and south Wales; many were feeding on small fish when stranded. These sailors are preyed upon by the violet sea-snail, which lives in the surface waters of the Altantic supported by a float of air-bubbles trapped in mucus. The two species are often stranded together. The much larger and more formidable Portuguese man-o'-war, with its poisonous tentacles, also appears occasionally on the Cornish coast. Goose barnacles from the South Atlantic sometimes arrive attached to driftwood while buoy-barnacles, which secrete their own spongy floats, are at times washed up in their thousands.

Whales, dolphins and porpoises are also liable to be stranded or washed ashore when dead, in varying stages of decay. Such animals are also occasionally seen off-shore. Pilot whales have accepted fish given by hand over the sides of boats; and in the late 1970s a dolphin spent eighteen months in Mount's Bay becoming extremely tame. 'Beaky', the name by which this delightful animal was known, was a common dolphin, the species most frequently stranded, but bottle-nosed, Risso's and white-sided dolphins have also been recorded — five of the latter species came ashore near Bude in 1971 but died before they could be returned to the sea. Rorqual and blue whales have come ashore in past centuries but the most frequent strandings are of pilot whales, cast up dead. In 1961 a large school of these animals would have been stranded in Gillan Creek had not local people and visitors driven them back into the sea. (The fish, including sharks, of the Cornish seas are considered to be outside the scope of this book.).

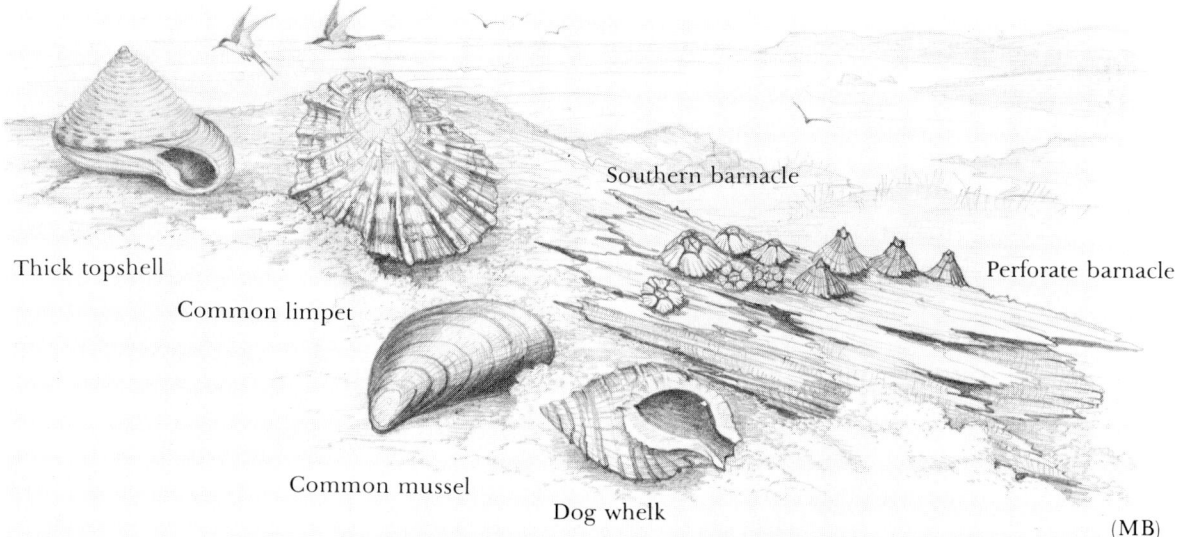

Thick topshell

Common limpet

Southern barnacle

Perforate barnacle

Common mussel

Dog whelk

(MB)

Nature Conservancy Council and the Marine Biological Association have graded beaches in Cornwall in terms of exposure, more than fifty being considered 'of primary importance.' There are more exposed beaches, exposed or moderately exposed beaches, and those which are sheltered or very sheltered.

Most of the *More Exposed Beaches* along the north coast are broken by rocks with densely packed beds of mussels; above these are southern barnacles, limpets, winkles and the purple dog-whelk. Where there are rock platforms pools are plentiful with usually common blenny, beadlet and snakelocks anemones, limpets and topshells with encrusting seaweeds and tubeworms. At the lowest tides, deeper pools, crevices and overhangs reveal crabs, shrimps, sea-squirts and sponges. Typical beaches of this kind are at Newquay and Strangles, north of Boscastle. On the barnacle-covered granite of Land's End and Cape Cornwall there is little plant life other than lichens; there are not many mussels. Seaweeds occur in exposed bays only where boulders provide some shelter; certain sea-urchins are also likely to be found — the common sea-urchin is abundant in shallow water, where collectors allow it to survive, but is seldom found between the tide-lines. Barnacles are dense on the exposed beaches (boulder and granite sand) of St Agnes in the Isles of Scilly.

Most of the *Intermediate Beaches (Exposed or Moderately Exposed)* have good, flat rock platforms where some shelter is to be found, particularly in pools. Animals associated with these rock-pools, some of which are encrusted with algae, include limpets, winkles, sponges, sea-urchins, worms and crustaceans. The zoning of seaweeds is not always clearly defined. Among a wide variety of such beaches is Trevone, where broken rocks and reefs provide both fully and partly exposed conditions. Many rocks are covered with mussels, limpets (including abundant china limpets) and barnacles. A population of the common hermit crab, often found in shallow pools, appears to have been destroyed by pollution at the time of the *Torrey Canyon* disaster. Duckpool, north of Bude, is a small bay, parts of which are very exposed, but shelter is provided by numerous rocky reefs; there is also some sandy shore and a small pebble beach. It is noteworthy because of exceptional colonies of the reef-building *Sabelleria* worm, whose sand-tubes are grouped in large masses — Duckpool is considered to be the best site in the British Isles for this species. Other examples of intermediate beaches are to be found near Looe and Falmouth, both with dense banks of seaweeds on the lower shore, and in the Battery Rocks area of Mount's Bay. Parts of the latter, where there is a good range of marine animals, are thick with barnacles. Porth Cressa, on St Mary's where there are granite rocks and gravelly sand, is an example from the Isles of Scilly.

On the *Sheltered and Very Sheltered Shores* of the south coast there are belts of seaweeds with, on the rocks, a profusion of animal life: sponges, hydroids, sea-slugs and sea-anemones. There are comparatively few mussels and, though barnacles tend to be sparse, both northern and southern species are present. The sandy shore of Camel's Cove, near Nare Head, is interrupted by reefs and boulders. There is a relatively sparse cover of seaweeds but mussels are plentiful on some of the more exposed reefs and, where cliffs face southwards, there is a profusion of southern barnacles. Porthnadler Bay, west of Looe, is relatively sheltered and provides many rock-pools for the appropriate animals, including marine bristle-worms which make their tubes with mucus, and the star sea-squirt which appears in a marvellous variety of colours. Both Helford River and St Mawes inlet, across the Carrick Roads, are very sheltered. Place Cove, part of the inlet where there is both sand and sheltered rocky shore with trees overhanging the beach, demonstrates particularly well the zoning of seaweeds and other forms of life. There are bristle-worms with tubes of mud projecting from the ground as well as burrowing bivalves, including the pod razor, striped Venus and thin tellin; the Darwin acorn barnacle, an Australian species which first appeared in England towards the end of the second World War, is also present.

Virtually no plants grow on sandy shores between the tide-lines, and the animals burrow into the sand and live below the surface when the tide goes out, in order to avoid exposure. These are the burrowing crustaceans, molluscs and worms as well as animals such as sandhoppers and sand-eels — food for wading birds, which feed along the edge of the sea as these submerged animals come to life again with the incoming tide. Stones often provide shelter for small porcelain crabs. The broad-clawed species *Porcellana platycheles* is found on many Cornish beaches; the long-clawed *P. longicornis* is also present in the middle and lower shore but is much less common. The common shore crab and the small showy spider crab also shelter under stones and may be found between the tide-lines. The edible crab is associated with rocky shores, and small specimens are quite common — large specimens live off-shore. Other crabs associated with rocky shores are the furrowed and hairy crab: both have a largely south western distribution.

The main threat to all this wildlife and to that of the oceans themselves is human disturbance (including over-use of resources and collecting), and pollution. This is a highly complex subject but few can be totally ignorant of the effects of pollution. On the beaches of Cornwall it is inescapable. Plastic bottles, oiled sea-birds, black lumps of oil which look like tar, and rubbish of almost indescribable variety lie along the strand-lines. Oiled birds are always present, not only when some tanker disaster has been reported. Shipping of all sizes constantly empties its tanks, depositing oil and rubbish into the sea. The *Torrey Canyon* wreck of 1967 is still the best known and best documented of marine calamities. Not only birds but crabs, whelks, small fish and almost every known form of plant and animal life were destroyed in the affected areas — as much, if not more, by the detergent used to break up the oil as by the oil itself which only damages organisms that it actually covers. Since the 1967 disaster Porthmear beach, St Eval, has been regularly monitored by Richard Pearce. The beach was heavily polluted with oil. Less than a week later it was 'cleaned' with concentrated detergent. The detergent was poured on to the beach when the tide was going out — not when it was coming in as biologists advised — so that it lay undiluted on the sand and rocks for up to ten hours. The effect was catastrophic. However, the green seaweeds of the foreshore began to recolonise the beach almost immediately. Brown seaweeds followed and spread before enough limpets had returned to the beach to control them by browsing. After a few years, however, the balance readjusted itself, but not until a complete generation of limpets (three or four years) had come and gone; where the first recolonising limpets came from is not known. Even so, green seaweeds, brown seaweeds and limpets reached a stable equilibrium within a period of six years and have remained in balance ever since. These were and still are the dominant forms of life on Porthmear beach. The pre-1967 community is not on record, so it cannot be stated with certainty that the life of the beach has returned to absolute normality. But that it should have 'returned from devastation to equilibrium' within six years shows the extraordinary vitality of seashore life. It would take decades for a comparable community of terrestrial plants and animals to recover to the same extent.

There are no marine or seashore reserves in Cornwall. Their establishment is fraught with difficulties involving ownership, public rights of access, fishing (and, by inference, collecting) rights and the virtual impossibility of establishing adequate physical control — the historical right to wrecks may even be involved. The desirability and importance of such reserves is unquestionable. It is as yet an unresolved problem. Given the goodwill of the owner of a stretch of foreshore, it would be possible to declare a reserve over some remote and inaccessible cove but nothing, except possibly kudos, would be gained as such areas are in effect nature reserves already. It may well be that effective seashore or marine reserves can only be set up by local authorities with strong government backing which, following the passage of the Wildlife and Countryside Act, may soon be forthcoming; but even then the problem will remain of controlling access — chemical as well as human — from the sea. Seashore conservation is a matter of immense importance. With the flowers of the coast, the life between the tides is perhaps the most typically Cornish element in the county's splendid wildlife scene.

Small tanker discharging oil at sea but close inshore. (FC)

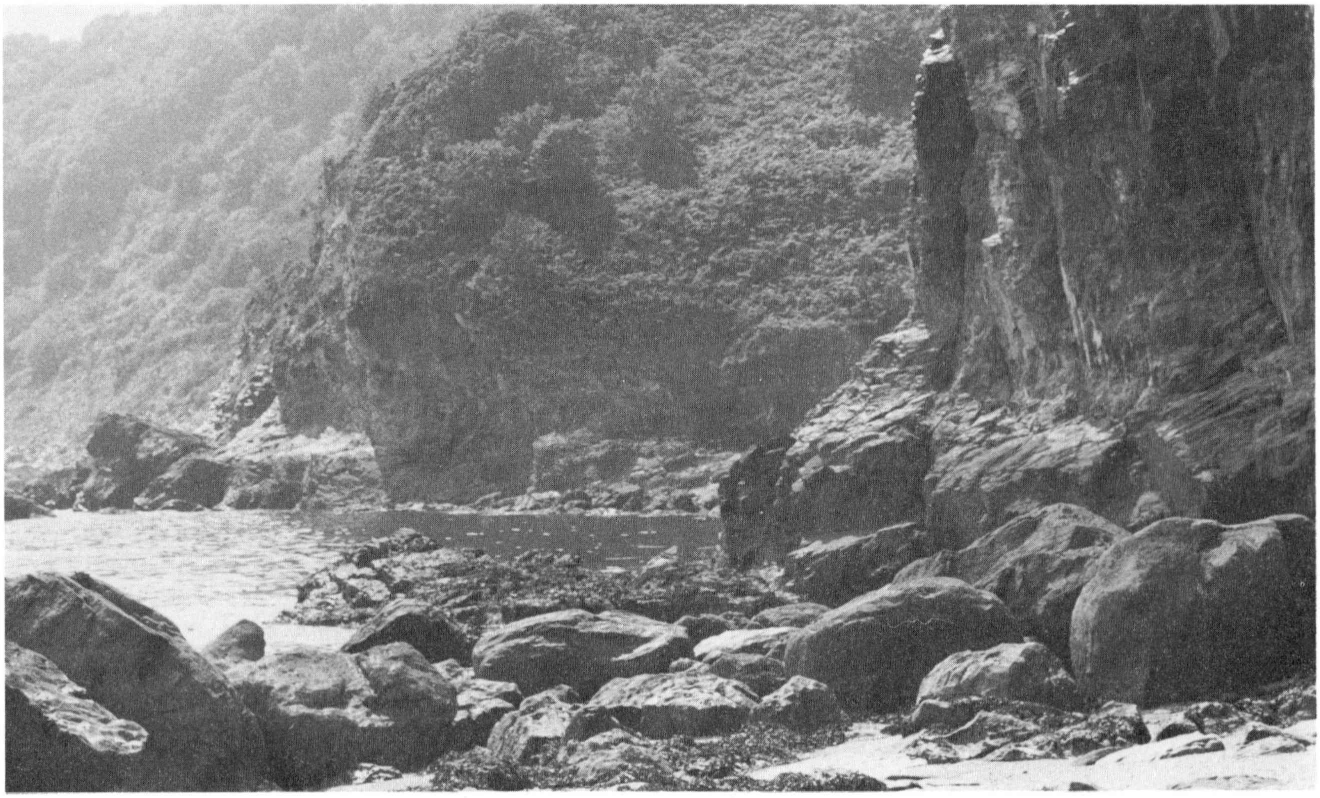

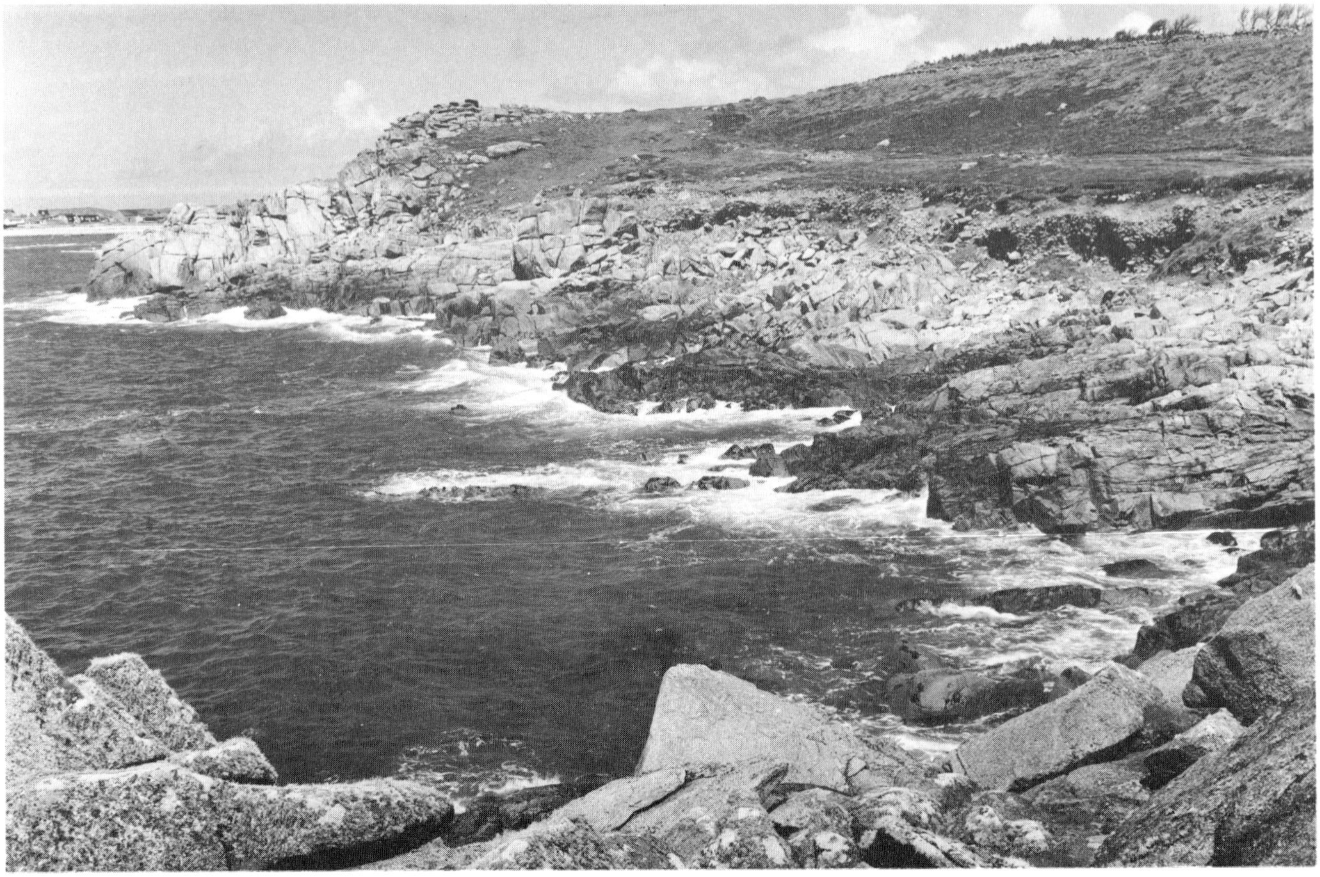

OPPOSITE ABOVE: 'Silver Mines' near Porthpean — a moderately
sheltered south coast beach; (CGB) BELOW: Peninnis Head, St Marys, an
exposed rocky shore on the Isles of Scilly; (NCC) ABOVE: a stony beach at
Perranuthnoe in Mounts Bay, (CW) and BELOW: Pentile Bay, a sheltered
sandy beach on Tresco in the Isles of Scilly. (NCC)

LEFT: Sabellaria reefs with encrusting barnacles and limpets above a rock pool on the north coast — massed sand tubes made by the sedentary Polychaete worm *Sabellaria alveolata;* (JB) RIGHT: Laver, the edible seaweed, above a rock pool, (JB) and BELOW: edge of a rock pool with Beadlet anemones, barnacles and limpets on an exposed north shore. (JB)

The Isles of Scilly

St Mary's from the causeway between St Agnes and Gugh. (FC)

The Isles of Scilly deserve a book to themselves rather than a short chapter. Though an integral part of Cornwall and the culmination of the county's granite spine, the islands constitute a special area of great natural history interest in spite of their limited size. The Trust, however, has not yet been much involved in the islands' conservation. With the exception of Tresco and smaller areas on St Mary's, which are leased or owned in freehold, the whole group is Duchy of Cornwall (*ie* Crown) property. There is strict control over development. The uninhabited islands and outlying rocks are under the guardianship of the Nature Conservancy Council, with restricted access, particularly during the breeding season of birds. Thus they are effectively nature reserves; Annet, where Manx shearwaters and storm petrels breed, is a declared nature reserve. There is a bird observatory on St Agnes.

The islands, only five of which are inhabited (St Mary's, Tresco, St Agnes with Gugh, St Martin's and Bryher), vary in size from St Mary's (two miles across), to the isolated rock pinnacles of Maiden Bower. In earlier geological ages the islands were part of the mainland. At the end of the last Ice Age, when the sea was some thirty feet lower than it is today, the group probably consisted of one large island with Annet, St Agnes and Western Rocks separated from it. Though never covered by the ice cap, soil at certain periods was frozen right down to the rocks below. At warmer times half-melted soil, rock fragments and gravel (together known as 'head'), were deposited in valleys by the process known as solifluction or 'soil creep'. Much of this has subsequently broken down into sand and gravel with massive blocks of underlying granite revealed by erosion as at Peninnis on St Mary's. The islands are thus largely composed of granite, 'head', sand and alluvial soil.

There is much evidence of early human habitation, with Stone Age axe-heads and flints, Megalithic pottery and Bronze Age burial grounds. Bones of both roe and red deer as well as other mammals have been found at Bants Carn on St Mary's and of numerous birds (including gannet, cormorant, white stork, razorbill and song thrush) in the pre-Roman Iron Age site at Nornour off St Martin's. The islands were inhabited right through the Roman period and until the turbulent 15th century, when raiding by outlaws and pirates, who found convenient refuges in Scilly, made life virtually impossible. Re-occupied in the 16th century, the human population reached 2,000 at the end of the 18th, remaining reasonably constant ever since. There were small populations on Samson, St Helen's and Tean before 1855. Some other islands were formerly used for pasturing sheep, burning kelp, collecting seaweeds used as a feed for domestic animals and as manure, collecting driftwood and as landing places for fishermen. The main preoccupations of the islanders were subsistence farming and fishing.

Today the economy depends largely upon flower-growing (with early potatoes included in most rotations) and tourism — many visitors come to the islands for the wildlife, particularly birds. Three-quarters of the human population live on St Mary's, about half of which is cultivated; there is relatively less cultivation on the other islands. The total area of farmed land is 1,600 acres (overall land area c4000 acres), of which flower-growing accounts for some 37 per cent, mostly in units of from five to eight acres. There is some dairy farming. Sheep, pigs and poultry are kept in small numbers. The largest single agricultural unit is on Tresco.

Agricultural expansion is limited by exposure and the rough surface of much of the land, particularly the extensive downs, which account for a considerable part of Tresco, Bryher, St Agnes and St Martin's. To flower-growers, the most dangerous element is the almost continuous wind, which brings with it salt from the sea — the rainfall (average 32 inches), like the temperature, is moderate. Although there is evidence of woodland in prehistoric times, there are few natural trees today; most of those now growing have been planted in clumps or belts to provide shelter or for amenity reasons. The Abbey Wood on Tresco, where there are fine Monterey pines, is the only wood of any size. Fields are enclosed either by granite hedges or by planted 'fences' of tamarisk, pittosporum, escallonia, euonymus or veronica: the familiar tidy 'squares' of the Scilly horticultural landscape. This contrasts splendidly with the rugged heaths

Yellow-horned poppy. (MB)

and downland, covered with gorse, heather and bracken. When the blooms have been harvested, bulb fields soon become overgrown and, for a while, wild flowers take their place — they are soon destroyed by cutting and cultivation. There are carpets of bluebells, an abundance of thrift near the sea and many colourful splashes of exotic mesembryanthemums. It is an enchanting scene, with sandy beaches, boulder-strewn shores and sheltered coves, as well as massive granite headlands and rock stacks exposed to the full force of the Atlantic gales.

Virtually all the main habitats are connected with the coast, and all are strongly influenced by proximity to the sea. There is the short turf of the upper shores and low hills; maritime heathland mainly in Tresco, Bryher and St Martin's; dunes and sandy grassland also mainly on Tresco and St Martin's. Areas of pasture land are of relatively small extent.

The Tresco woods and Abbey gardens, from which several sub-tropical species have escaped into the open (agapanthus lily, tree lupin and New Zealand flax, for example), together with the shelter-belts and wind-breaks, are essentially artificial habitats, but they introduce to the islands areas where woodland conditions prevail and certain woodland birds can find food, protection and nesting places. There are few streams, but the Great and Abbey Pools of Tresco, freshwater lakes impounded behind accumulations of blown sand, attract numerous wildfowl and other water birds. The main seashore and inter-tidal habitats are all represented from extremely exposed rocky shores, where the flora and fauna are similar to that of the Land's End granite, to sheltered sandy beaches protected by rocks and boulders. The seashore life benefits greatly from the equable climate and the islands' continued freedom from pollution; the virtual absence of rivers means that little brackish freshwater, inimical to many marine species, enters the sea.

Taking the flora and fauna as a whole, the isolation of the islands has led to a reduction in the number of species (by comparison with the mainland), and the development of variety. Even so, the climate assures an abundant and spectacular display of wild flowers, though several familiar plants are absent. Among these are the heath spotted orchid, cross-leaved heath which grows on wet heathland throughout mainland Cornwall, and the thistle-like sawwort. Kidney vetch seems to have become extinct during the present century. By contrast, several species which are rare on

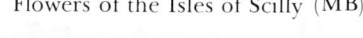

Flowers of the Isles of Scilly (MB)

Sea storksbill   Dwarf pansy

the mainland grow well on the islands: sea kale, white mignonette and yellow-horned poppy, for example. The uncommon fern, lanceolate spleenwort, grows well on granite hedges. Two plants which normally appear only in woodland thrive in the open on the islands: butcher's broom, which produces scarlet berries in the autumn, and wood spurge which grows among the heather. The rare seaside buttercup is to be found in shallow pools on St Agnes and Bryher. A number of interesting plants grow on the extensive dunes on Tresco and in other sandy areas; among them are orange birdsfoot with red-veined yellow flowers, dwarf pansy which occurs only on the Channel Islands and Scilly, sea storksbill and Portland spurge. The prickly-fruited Scilly buttercup, rosy garlic, Babington's leek, three-cornered leek and the Bermuda buttercup are all common in bulb fields, sometimes to the point of becoming pests. Both common and narrow-leaved eel-grass grow submerged in the shallow seas between the islands, as they do on the tidal mud-flats of certain mainland estuaries.

Several familiar butterflies are missing. There are no graylings, skippers, orange-tips or commas. The only representative of the 'browns' is a Scilly subspecies of the meadow brown,

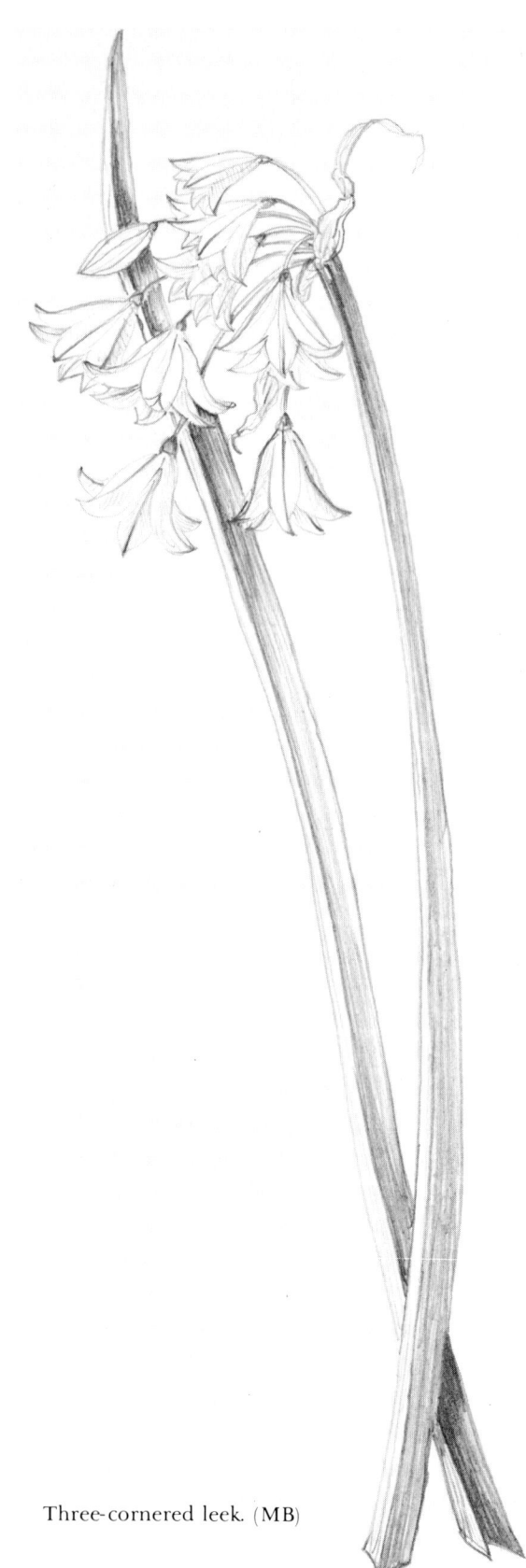

Three-cornered leek. (MB)

Maniola jurtina splendida, found also in Ireland and on the Scottish islands. There is also an island variety of the common blue which is restricted to Tean. The splendid monarch or milkweed butterfly, which has a wing-span of nearly four inches and is an occasional visitor from North America, is more likely to be seen on the Scillies than anywhere else in Britain. No bumble-bee, not already established in mainland Cornwall, has been found on islands, though some species may be absent. There is a Scilly race of the moss carder bee, *Bombus muscorum scyllionius,* which appears to be declining; it is no longer found on St Mary's and is now restricted to uncultivated areas. Reasonably plentiful are the large earth bumble-bee and, wherever there is heather, the heath bumble-bee; the vestal cuckoo bee is also well distributed. Of the eleven true grasshoppers on the British list, only the common field grasshopper occurs on Scilly, and of the ten bush-crickets only two are known: the great green and the grey bush-crickets. There are no groundhoppers. Only one of the three native British cockroaches is found on the islands: the lesser cockroach. Of introduced cockroaches, which are more numerous than native species, there is only the common cockroach. The smooth stick-insect is found on Tresco and an island off the coast of Kerry in south-west Ireland, but nowhere else in the British Isles; the prickly stick-insect also occurs on Tresco, but nowhere else except for a single site in south Devon. Both are natives of New Zealand and were probably introduced accidentally on imported plants.

There are no toads or snakes; frogs occur on St Mary's but not on other islands. Several familiar mammals are absent: pigmy shrew, water shrew, short-tailed vole, dormouse, stoat, weasel, fox, badger and otter. There were no squirrels until 1968, when the grey was first noted on Tresco — though grey squirrels swim well, it can hardly have reached the island without human help. Bat records from Scilly are scarce, none being of recent date, but the pipistrelle was present and apparently quite plentiful early in the present century.

The Scilly or lesser white-toothed shrew is not unique to to the islands; it is widely distributed in continental Europe but does not occur on the British mainland. Its main characteristics are a bristly tail, prominent ears and the absence of red pigment in its teeth. While found occasionally in bracken and on grassland, it is most often seen on stony beaches where it feeds on sandhoppers and insects; it escapes from the brown rat, its main enemy, by moving about between stones where it cannot be followed. Another mammal of particular Scilly interest is the grey seal. Most Cornish seals breed in caves; on the islands they breed above high-water mark on remote boulder beaches and shelving slabs of rock. The main breeding areas are on Western Rocks and islets such as Gorregan and Melledgan; a few also breed on the northern islands and Eastern Rocks but evidently find this part of the achipelago less attractive. The total population is about 120 animals; some 40 pups are born annually in the autumn. There seems to be a regular interchange of pups between mainland Cornwall, the Isles of Scilly and Pembrokeshire, with constant movement between these three areas.

The position regarding birds is strongly influenced by the extreme south-westerly and oceanic situation. The number of *breeding* species is markedly less than on the mainland but the number of species *recorded* is greater. Of 409 bird species so far recorded in the county as a whole *(Cornwall Bird-Watching and Preservation Society's Annual Report for 1980)* 40 have been recorded on Scilly but not on mainland Cornwall, as opposed to 33 on the mainland which have not been seen on the islands. The first category includes many birds blown off course from across the Atlantic and 'accidentals' from south-west Europe: scarlet tanager, bobwhite, American purple gallinule, semi-palmated plover and blue-cheeked bee-eater, for example. Among mainland species never recorded in Scilly are such familiar birds as the dipper, lesser spotted woodpecker and nuthatch — ornithologists interested in this fascinating subject should consult the report. The only bird of prey to breed on the islands is the kestrel; others are seen only occasionally. No owls breed. Few corvids are in evidence, and there are no breeding ravens, jackdaws, magpies or rooks. There are no woodpeckers and no tree-creepers. Skylark, wren, song thrush, blackbird, dunnock and house sparrow are probably the most common of the smaller birds. Great and blue tits have nested on Tresco and certain other islands. Ringed plover, which favours shingle beaches, and oystercatcher are the most abundant wading birds. Both nest on inhabited as well as on uninhabited islands.

In spite of the rarities, it is the seabirds which are the true glory of the Isles of Scilly. Not many birds breed on the inhabited islands but there are important breeding colonies on many of the outlying stacks and uninhabited islands. Annet is one of the few places in southern Britain where Manx shearwaters and storm petrels breed; both species nest in holes or burrows. Razorbills also nest there, but their largest colony and that of guillemots is on the northern stack of Men-a-vaur. Large numbers of kittiwakes also nest on Men-a-vaur as well as on a low earth cliff on St Helen's — these most graceful gulls usually nest on high and inaccessible ledges. Shags, cormorants, several species of gull, common terns (which do not breed on the mainland of Cornwall) and, irregularly, fulmars also nest on the islands. Though there may be some paucity of species, there is on the Isles of Scilly a remarkable diversity of interest to suit the taste of every naturalist.

Watching a rare bird on the Isles of Scilly. (FC)

ABOVE: A carpet of thrift below Annet Head; (NCC) OPPOSITE ABOVE:
Peninnis Head, St Mary's — a typical island scene, (NCC) and BELOW: an
exposed rocky shore on Annet, the bird sanctuary island. (NCC)

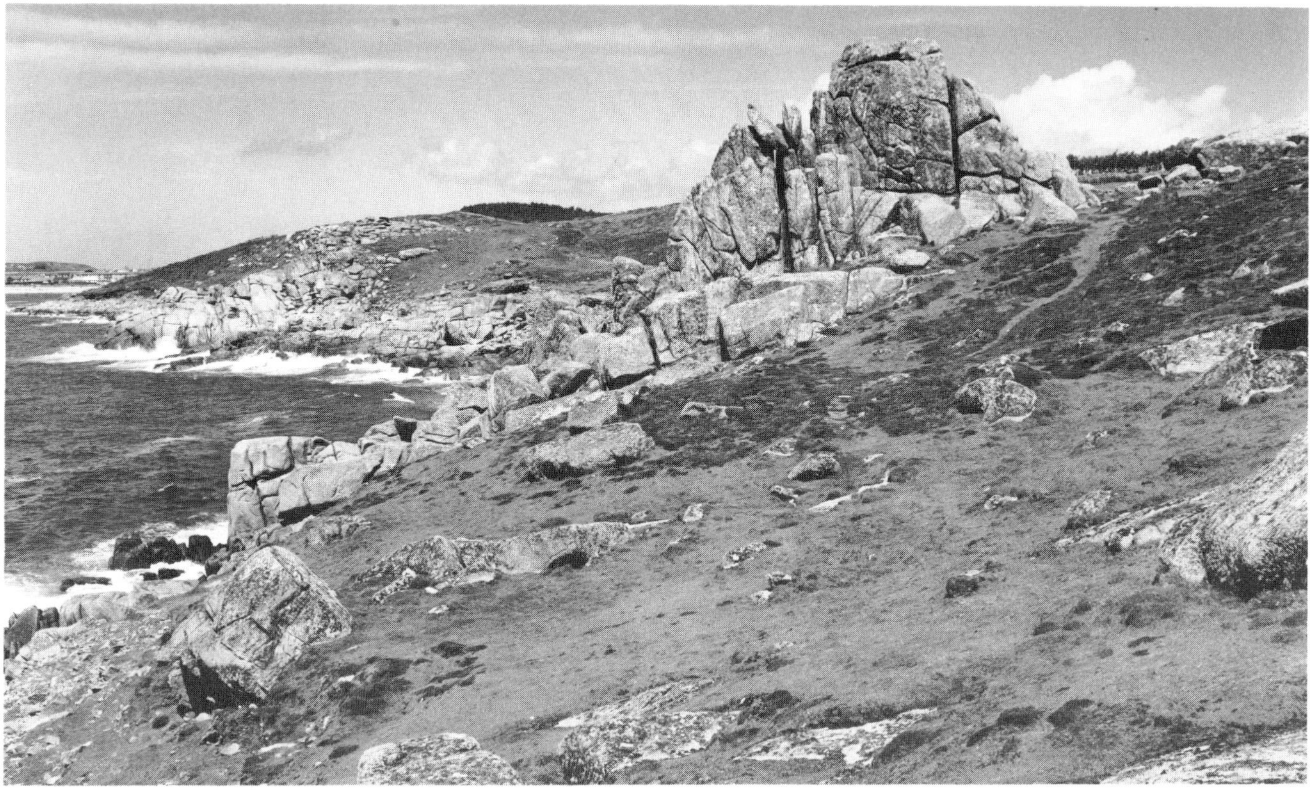

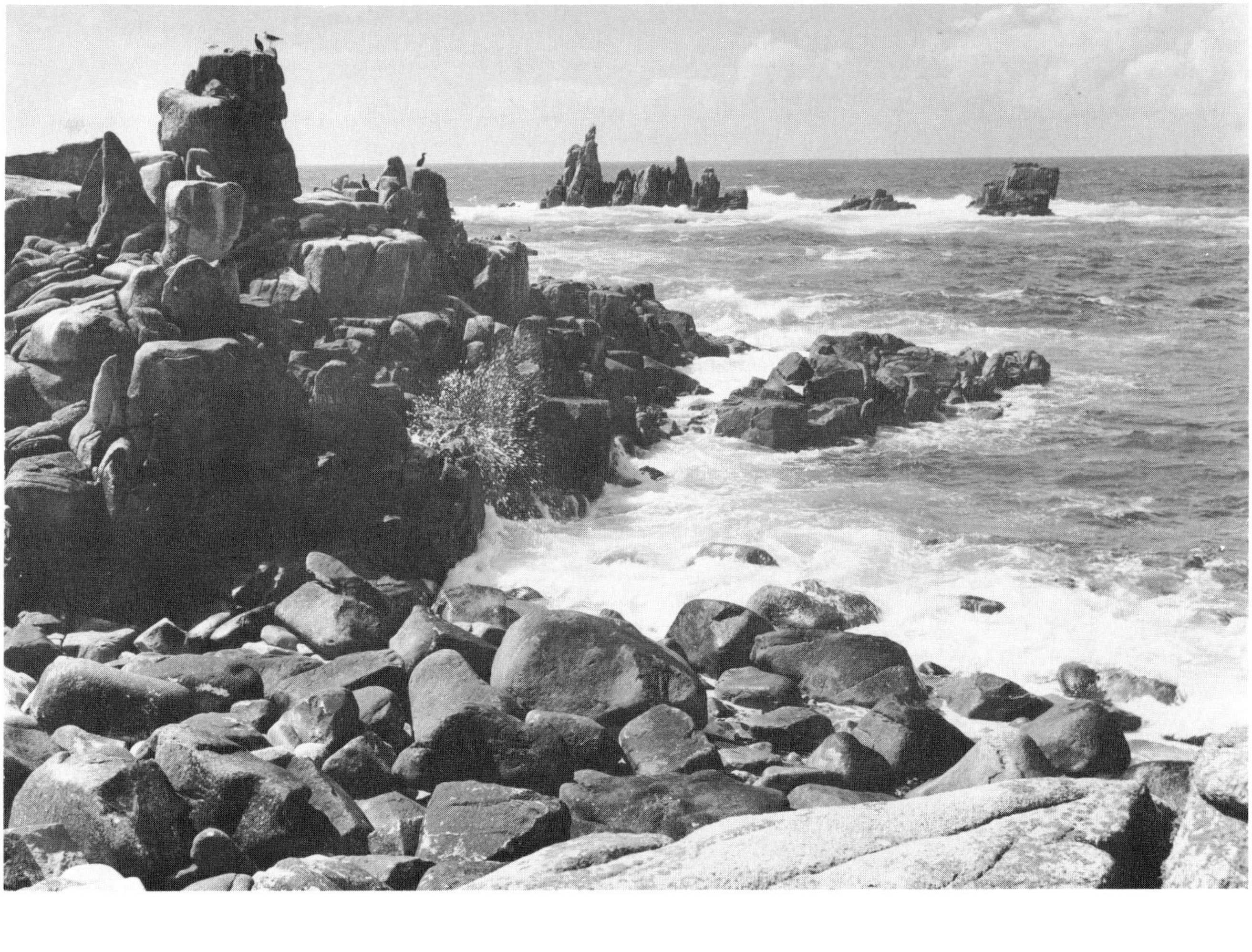

ABOVE: Sandy grassland — Pentle Bay, Tresco, (NCC) and BELOW: Rushy
Bay, Bryher — locality for the Dwarf Pansy. (NCC) OPPOSITE: Porth
Hellick Pool on St Mary's. (NCC)

LEFT: Shelter trees and bulbfields weeds on Wingletang Downs, St Agnes; (NCC) RIGHT: Great Pool, Tresco, with Abbey Wood in the background, (NCC) and BELOW: Bulb square with Pittosporum and stone walls.

(F.E. Gibson)

Conserving the Nature of Cornwall

Jackdaws in flight around old mine tip buildings. (FC)

The most important single element in the conservation of nature in Cornwall is the National Trust. Founded in 1895 to act as trustee on behalf of the nation for places of historic interest and natural beauty, the objective was to halt the destruction which industrialisation was already bringing about and, at the same time, to give people access to the countryside subject to the demands of good farming, forestry and the protection of nature. In 1907 parliament conferred upon the National Trust the unique power to declare land inalienable, a condition which now applies to most of its properties, and enjoined it to preserve 'so far as is practicable' the plant and animal life on its lands. All National Trust lands are protected against development, but most must be made to yield a satisfactory economic return, provided this does not interfere with the primary objective of preservation. Properties vary greatly in importance in natural history terms, but some are managed in such a way that the nature reserve aspect is allowed to play the dominant role. Today this is usually achieved by lease to the Nature Conservancy Council or to the appropriate county naturalists' or conservation trust. In Cornwall, a large part of the Lizard National Nature Reserve is on National Trust land; two Lizard reserves and three areas elsewhere are leased to the Cornwall Naturalists' Trust: two sections of the Fal estuary and Peter's Wood near Boscastle.

These areas form only a small part of the National Trust properties in Cornwall, which now amount to 17,052 acres owned, including approximately 102 miles of coast, with an additional 7,030 acres and 15 miles of coast protected by covenant. It is these properties which are of such outstanding importance, as they include most of the finest stretches of Cornwall's magnificent coastline. The coast footpath passes through these properties; it does no harm to the natural environment and introduces many people to the glorious wild flowers and bird life as well as to the incomparable sea views. One of the National Trust's earliest acquisitions was Barras Nose on Tintagel Head, secured by public subscription in 1897. Among other areas abounding in wildlife are Morwenstow cliffs, Dizzard Point with a large section of the stunted oak forest, Cambeak with the Strangles and High Cliff, Boscastle cliffs with Forrabury Common and the Valency valley, Pentire Point and the Rumps, St Agnes Head, the six mile stretch, largely of coastal heathland, between Portreath and Godrevy, Zennor Head and Rosemergy cliffs in Penwith, Chapel Carn Brea, important sections of the Lizard coast and of Helford River, parts of the Fal estuary, the Dodman, Nare Head and Veryan Bay, Erth Island and salt-marsh areas in the Lynher valley.

Inland there are National Trust woods and parkland at Lanhydrock, Trelissick and Cotehele, as well as 174 acres of moorland on Rough Tor — the only part of Bodmin Moor which is properly protected. As well as all this, the National Trust has given constant support and encouragement to the Naturalists' Trust and is the co-operative landlord of the latter's comfortable offices at Trelissick. During a period of crisis in the Trust's affairs, the then land agent even took over for several months the arduous job of honorary secretary. Such National Trust properties as Trelissick and Lanhydrock would not be what they now are were it not for the care and good husbandry of former landlords. And there still are many fine estates in private ownership where everything possible is done to preserve the best features of the countryside; the Tregothnan estates, Bocconoc and Chyverton are particularly good examples.

Though its status was confirmed by Act of Parliament, the National Trust is a completely independent body supported by voluntary finance provided by its members, benefactors and those who make use of what it has to offer. In most respects Britain has lagged behind the rest of the world in legislating to protect the countryside and the wild creatures that live in it. Until quite recently birds and other animals have been seen only in terms of sport, with close seasons for such species as grouse, pheasant and deer, and no mercy shown to predators — other than man. Not for us the great national parks of other lands: 'the pleasuring grounds for the people' first established in the USA in 1872 where the 'pleasure' was (and still is) in seeing wild birds, other animals and vegetation in their own magnificent natural surroundings. There are problems in these small islands but, with hindsight, a great deal more could and should have been done. Even today, conservation seems to have an appallingly low priority in the view of government, the Countryside Act of 1968 notwithstanding; this acknowledged the role of government (including every individual minister, department and public body) in conserving the fauna, flora and geological features of the countryside.

Involvement of the British government in nature conservation began with the National Parks and Countryside Act of 1949, which resulted in the National Parks (later Countryside) Commission and the Nature Conservancy. In Cornwall there are no national parks but there are Areas of Oustanding Natural Beauty, designated under the same act and of comparable status in the sense that these areas enjoy planning control stricter than elsewhere in the countryside. The Cornwall AONBs (with associated Areas of Great Landscape Value) include almost the whole coast, a large part of Penwith, the Lizard peninsula, Fowey valley and the whole of Bodmin Moor — some 350,000 acres in all. About two-thirds of the coast is also designated, under the same act, as Heritage Coast which is an even stronger concept. It is Cornwall County Council policy, unequivocally stated in the County Structure Plan (1979) and approved by the Ministry of the Environment, to assure protection of these areas, though pressures from such interests as tourism and mining can be formidable. The Act also gives local authorities power to establish *local* nature reserves. Nothing much has yet been done except for the small Bude marshes reserve in North Cornwall. But other district councils are showing interest; Par beach with Polmear lake (Restormel DC) and Kilminorth wood near West Looe (Caradon DC) are likely to become reserves in the near future.

The Nature Conservancy (now Nature Conservancy Council, part of the National Environment Research Council) was set up to advise government and local authorities on nature conservation, to establish and manage National Nature Reserves (as, for example, on the Lizard) and to carry out research. NCC also has the duty of providing local planning authorities with details of areas or sites of special scientific interest — SSSIs as they are known. Local authorities have a comparable duty to consult NCC and take account of the scientific interest when considering planning applications likely to affect SSSIs; unfortunately, however, changes in agricultural practice and afforestation are not controlled by planning law. There are some seventy SSSIs in Cornwall, and in spite of the limitations on their control this is an extremely useful provision. Much of the coast is covered, as is Bodmin Moor north of the A30 trunk road. Dune areas, woods such as Draynes and Trenowth, worthwhile bogs, salt-marshes and even some old quarries are listed. There has been a resident NCC officer in Cornwall since 1967, and liaison with the Trust is close.

The two Protection of Birds Acts (1954 and 1967) are perhaps the only really effective pieces of legislation on the statute book. They not only protect certain vulnerable species, but actually accept the principle that a wild creature can have rights to share a common heritage with man. Of particular interest to Cornwall is the fact that the drafting and passage through parliament of the 1954 bill owed much to the late F.H. Hayman, MP for Camborne at the time. Mr and Mrs Hayman, who were active founder members of the Trust, are remembered today by the Hayman Reserve, Park Hoskyn, a wooded valley near St Agnes.

The Conservation of Wild Creatures and Wild Plants Act, 1975, was primarily concerned with the total protection of a small number of rare species, including twenty-one plants and animals such as the large blue butterfly which was already virtually extinct. It also prohibited the uprooting of any wild plant without permission from the landowner, a condition which restricted the value of the ban. Even so, in 1981, two men were successfully prosecuted for digging out a number of Cornish heath plants on the Lizard. The Badgers Act, 1973, would have given effective protection to badgers which, through the ages, have suffered greatly from man's brutality, but for the bovine tuberculosis scare — a matter of intimate concern in Cornwall. However, the Act had to be amended to permit gassing by 'qualified persons in infected areas' — no one can now describe the badger as a well-protected animal.

The Wildlife and Countryside Act 1981, like so much earlier legislation, fails to deal with many of the problems that concern wildlife conservation bodies throughout the country: snaring of wild animals, still prevalent in Cornwall, and the destruction of valuable hedges, for example. Nor does it give adequate protection to more than a handful of SSSIs or bring agriculture within the orbit of planning control. However, it does make provision for the establishment of marine nature reserves, which may extend from the strand-line to the limits of Britain' territorial waters, and it should also relieve some of the pressures on the countryside — the first of these provisions is of major importance to Cornwall. Unfortunately the Act comes when government funds are minimal for this or any other purpose. Above all else, more cash is required for wildlife conservation, particularly so in a poor county like Cornwall.

From 1602 when Richard Carew published the first lists of plants, birds and other animals in his *Survey of Cornwall*, knowledge of the county's natural history has grown steadily. The details of this progress in relation to plants and birds have been given in Davey's *Flora* and in a paper on *Ornithology in Cornwall* by R.D. Penhallurick in the 1980 (Jubilee issue) of the *Report of the Cornwall Bird Watching and Preservation Society*. There are a few distinct milestones. The Cambridge botanist, John Ray, 'the father of natural history in this country' according to *The Dictionary of National Biography*, visited Cornwall twice in the 1660s. Colonel George Montagu, author of *The Ornithological Dictionary* (1802), was stationed in the County during the 1790s, when he made notes on the Dartford warbler which was not then uncommon. Cornishmen and those whose home was in Cornwall were not far behind with Walter Moyle (1672-1721), whose collection and notes were lost in a fire, and Thomas Tonkin whose *Natural History of Cornwall*, begun in 1700, is still in manuscript in the county museum. The better known William Borlase, vicar of Ludgvan for many years, was more concerned with minerals and antiquities than with natural history as we now understand it; nevertheless his *Natural History of Cornwall* (1758 and recently republished) was a valuable contribution to knowledge.

By the beginning of the 19th century, interest in the wildlife of the countryside was growing, even if the emphasis was on collecting rather than conserving. Again a few names stand out: Jonathan Couch (1789-1870), many of whose observations were published in the journals of the Royal Institute of Cornwall (founded in 1818); E.H. Rodd (1810-1880), Cornwall's leading ornithologist, whose *Birds of Cornwall and the Isles of Scilly* was published posthumously; W.J. Hooker, the famous director of Kew Gardens; C.A. Johns, a master at Helston grammar school, who first described the flora of the Lizard and whose *Flowers of the Field* (1853) is still in print. *The*

Victoria County History (Cornwall volumes 1906), with contributions from Hamilton Davey on the flora and Dr James Clark on birds, is still a useful source of information. The works of J.C. Tregarthen, such as *Wild Life at the Land's End* (1904), though popular, are well worth reading. Davey's *Flora of Cornwall* (1909) and Colonel B.H. Ryve's *Bird Life in Cornwall* have become standard works of reference, the former now replaced by L.J. Margetts and R.W. David's *A Review of the Cornish Flora* (1981); other modern books are among those listed in the bibliography.

The Royal Institute of Cornwall, though primarily orientated towards historical and antiquarian interests, gave and still gives a valuable impetus to natural history studies, through its collection in the county museum and its journal. There have been other societies more directly concerned with nature including, in 1865, a Cornwall Natural History Society. Most such societies have been local and somewhat ephemeral, the first being The Natural and Antiquarian Society of Penzance, founded in 1839. Among such societies still active are the Lizard Field Club and Camborne-Redruth Natural History Society. There is also the Cornwall Bird-Watching and Preservation Society, a flourishing and progressive body concerned with every facet of ornithology and bird lore. Founded in 1931, largely through the efforts of Colonel Ryves, the society has developed steadily, its detailed annual reports essential reading for ornithologists. It either owns or manages the important Walmsley Sanctuary in the Amble marshes, Trethias island in Treyarnon Bay and part of Stithians reservoir; it holds the shooting rights, which are not exercised, over Hayle estuary and Restronguet creek, the latter threatened by impending mining operations. The Society is concerned in the management of several observation hides, mostly put up by the South West Water Authority, and works in close co-operation with the Trust.

Although this is the local background upon which the Cornwall Naturalists' Trust was established, a naturalists' or conservation trust is something different from a natural history society. It is not primarily concerned with the interests of its members. It is their function to serve the Trust (with expertise, hard work or cash) and help it achieve its aims — not the other way round. The aim is to benefit the area or county for which the Trust is responsible. Its business is conservation, an aspect of planned land use and the *wise* use of resources. The ideal of conservation, in biological terms, is maintenance of the optimum flow of energy intake from the sun, and its output through the growth and activity of plants and animals. This is best achieved through the natural succession of plants in any given habitat, and the animals which have adapted themselves to this vegetation. Of course, the whole countryside cannot be considered or managed in these terms, but it is necessary that part of it should be. This, therefore, is what the trust idea is all about. It cannot be achieved without detailed knowledge of the plants and animals and their distribution: hence the emphasis on study and recording — in addition to the Trust's records, in Cornwall this is also undertaken by the Institute of Cornish Studies — as well as on establishing and managing reserves and other protected areas. Again a trust cannot serve the county unless it is prepared to work closely with the authorities and make its expertise available. In return it needs to earn the right to be accepted and trusted by the authorities as a reliable adviser. How did the movement come about and where does Cornwall fit in?

The Society for the Promotion of Nature Reserves (now the Royal Society for Nature Conservation, or RSNC) was founded in 1912 by Charles Rothschild, who drew up a long list of scientifically important sites throughout the British Isles — most of these still are SSSIs. A small number of nature reserves was acquired. Then, in 1926, the Norfolk Naturalists' Trust was established, to be followed by Yorkshire (1946) and then Lincolnshire. By 1958 there were seven trusts who jointly approached RSNC and invited it to take over the role of promoting and sponsoring the movement at county level. This was the take-off point.

During the next few years most other counties, including Cornwall (1962), followed suit. RSNC provides trusts with a means of association and has set up a committee to which all trusts nominate members. It represents their interests at national levels as, for example, during the passage of the Wildlife and Countryside Bill. It secures funds for the various trusts from such agencies as the Nuffield Foundation and World Wildlife Fund. It has welded the trusts into a respected and cohesive movement without interfering with their independence. In 1958 the seven trusts and RSNC together owned or managed 34 reserves covering 6,500 acres. By 1970 there were more than 400 reserves covering 35,000 acres. Today there are 42 trusts with a membership of 137,000; there are 1,280 reserves, covering 109,000 acres.

Christopher Cadbury, president of the RSNC since 1962, has described the aims of the movement and of the individual trusts as to assure the greatest possible variety of wildlife in the widest possible variety of ecological habitats; to see that the countryside is managed in the interests of its flora and fauna as well as of man; to accept that the needs of wildlife have to be reconciled with the necessity for food production and human recreation; to assure that the owners, users and planners of land are aware of the need for conservation so that they do not unwittingly destroy what should be preserved; to encourage young people to learn something about nature conservation and to participate in the work of the trusts; to establish as close a relationship as possible with local authorities and other public bodies as well as unofficial organisations (such as the National Trust, National Farmers' Union and the Council for the Protection of Rural England) with relevant interests.

The Cornwall Naturalists' Trust, which has always worked along these lines, was founded in May 1962, following a public meeting chaired by the late Colonel W. E. Almond in the County Museum at Truro, the scene of every annual general meeting since then and of many other Trust occasions. Officers were appointed. The county was divided into regions, each with a responsible naturalist in charge — initially the units were the eight botanical regions used by Davey in his *Flora* but, following the re-organisation of local government in 1974, the district council areas were adopted for easier liaison with the planning authorities. Species recorders and a sites recorder were appointed. Field meetings, some of which took the form of working parties or recording sessions, were held in different parts of the county. Reserves have gradually accrued by purchase, lease, agreement with landowners and bequest; the first was Hawke's Wood, near Wadebridge, leased by the late Miss Sewart, who subsequently presented the freehold. Other early reserves were secured by lease from the National Trust and the Duchy of Cornwall. The policy regarding reserves is to secure examples of every major habitat in as many different parts of the county as possible. Though much progress has been made, this objective is a long way from complete fulfilment.

Although there was a good number of SSSIs in Cornwall, and these included a majority of the county's most important areas in national terms, they by no means covered every area of real naturalist interest deserving or needing protection. The Trust therefore decided to initiate its own system of Conservation Sites — sites of scientific interest, important in local terms. This involved making an outline survey of the county and pooling knowledge from the different regions. To begin with, only brief descriptive notes could be made. More detailed ecological classifications have followed.

When the first list of Conservation Sites was ready, card-indexed and marked on ordnance survey maps, it was presented to the County Planning Officer as a preliminary overall picture of sites where the flora and/or fauna were of sufficient interest to merit conservation, even though not designated as SSSIs. Additions and amendments have been made as the county has become better known. Meanwhile, in the late 1970s, the Trust appointed its first full-time conservation officer; and the County Council was seeking more and more information, virtually amounting to a complete review of the Conservation Sites. The Conservation Officer, with a multitude of other duties, from reserve management plans to bolstering up district committees and attending planning enquiries, was no more in a position to provide all that was required than the amateur sites recorder had been. Fortunately, government's 'special temporary employment programme' provided the answer. A team of four graduates was engaged and has worked for more than a year checking, surveying and recording. There is now a considerable volume of ecological information on sites throughout the county available to both Trust and planning authorities.

In practice, and thanks to the remarkably cooperative attitude of Cornwall's planning officers, the arrangement with the County Council works extremely well. The Trust is informed whenever a planning application is made that refers to any of the Conservation Sites. A short report is then submitted. Although the Trust has no more statutory rights than any member of the public, the arrangement works as a thoroughly effective 'early warning' system. The ecological interest is never overlooked or ignored in areas where this is important. The Trust involves itself only in matters directly concerned with biological conservation and, as a result, has earned the reputation of being a responsible body whose views are worth listening to. It has also been able to assist with the overall planning of several potential development areas.

Probably the first overt sign of the Trust's acceptance by the establishment of Cornwall was the symposium on 'conservation in relation to land-use planning' organised with the support of the county planning office and Nature Conservancy Council, and held in County Hall in Truro on 12 May 1970 — European Conservation Year. Every local authority above parish level was invited and most took part, as well as representatives of the Ministries of Agriculture and of Housing, water authorities, Forestry Commission, National Trust, Duchy of Cornwall, National Farmers' Union, china clay and mining industries and many other public and private bodies. It was addressed by, among others, the chairman of NCC, Lord Howick of Glendale, who gave the concluding address. Lord Howick emphasised the importance of co-operation between conservation interests, planning authorities and those with a stake in the land: the interests of farmer, forester, business man and conservationist need not be incompatible, he said; 'somehow I manage to be all of these.' Since then a symposium has become a regular feature of the Trust's annual programme. Subjects have included the impact of tourism, mining, agriculture and forestry upon the countryside, hedgerow and road verge maintenance, and conservation education.

The 1970 symposium took place while the Trust was still a completely volunteer society without paid staff and without premises. Since then a part-time Administrator (1974) and a full-time Conservation Officer (1979) have been appointed — neither appointment would have been possible without outside financial help, particularly a grant from NCC to cover the Conservation Officer's salary for the first few years. Admirable premises at Trelissick have been rented from the National Trust. Gone are the days when anyone needing to contact the Trust might have to search Cornwall for a doctor in his surgery or a schoolmaster in class. But grants will not continue for ever; the Trust is badly in need of reliable new sources of finance and a larger membership.

Despite such problems, the Trust is being taken seriously. Farmers are beginning to seek advice. Developers, miners and oil men have at least come to realise that they can discuss proposals and that they will be given reasonable advice. The Trust, at the time of writing, is acting as conservation adviser to a 'hot rocks' experiment on underground sources of energy near Penryn. It has played its part in developing public awareness of the need for environmental protection — adding its voice to the increasing impact made by books, television and radio — and this should soon be materially increased by a travelling exhibition, produced with financial aid from the Carnegie Trust and the Countryside Commission. Contact with schools is gradually being extended, particularly around Liskeard and Camborne, where good educational use is already being made of the Pendarves reserve.

The Trust is also the active patron in Cornwall of WATCH, a national organisation aimed at involving children in conservation and teaching them something about wildlife. In many different ways the Trust provides a valuable conservation service to the county, and this is as it should be.

All this may seem to present an excessively rosy picture of the state of the county in conservation terms. The clifflands are well protected, thanks largely to the National Trust, and many more interesting beaches are too rocky or too inaccessible to be much disturbed by people. But sand-dunes are being destroyed, and arable is replacing natural grassland almost to the edge of many cliffs. Farming practice is changing the face of the countryside, even if more slowly than in many other parts of Britain. There is little heather left on Bodmin Moor, which is being threatened by both mining and new areas of afforestation — not that the latter are likely to be extensive. Ancient oak woods are being destroyed, and it will be a long time before the new plantations, with their rotations established, reach the stage that should make them an adequate replacement. Old mines are being re-opened and new areas prospected; china clay developments are being extended — on the credit site is the vegetation cover now taking over many old tips. Tourist pressure is increasing, and more land has to be turned into caravan parks, chalet camps and camping sites to cater for the growing number of visitors. Towns and villages expand and spread into the countryside. Land which serves as a haven for wildlife seems constantly to be subject to reclamation and development. Rivers and water courses generally are under pressure, because of increasing use of reservoirs for recreation, which may be inimical to bird life, an excessive number of fishermen and the demands of bank and verge management. Chemical pollution is an ever-present menace in many habitats.

It sometimes seems that the Trust and other conservation bodies can make little headway against such pressures. The damage is done piecemeal. There is no master plan among the agencies of destruction. And, with the best will in the world, the planning authorities are often powerless to resist what is presented as being of overwhelming human benefit, ephemeral though that may be. Man is part of the living landscape and cannot be totally divorced from the natural world without losing part of himself. 'Conservation is concerned ultimately with relationships' as Richard Mabey writes in *The Common Ground*, 'between man and nature and man and man;' and, one might add, a host of natural organisms which depend upon each other.

Cornwall is a beautiful county:

    'Where the ocean breaks
    On the purple strand
    And the hurricane shakes
    The solid land.' (Thomas Hardy)

It is worth great effort and some sacrifice, particularly in terms of financial rewards, to assure survival of its remarkably varied wildlife.

Grey seals on the rocks — peregrine above. (FC)

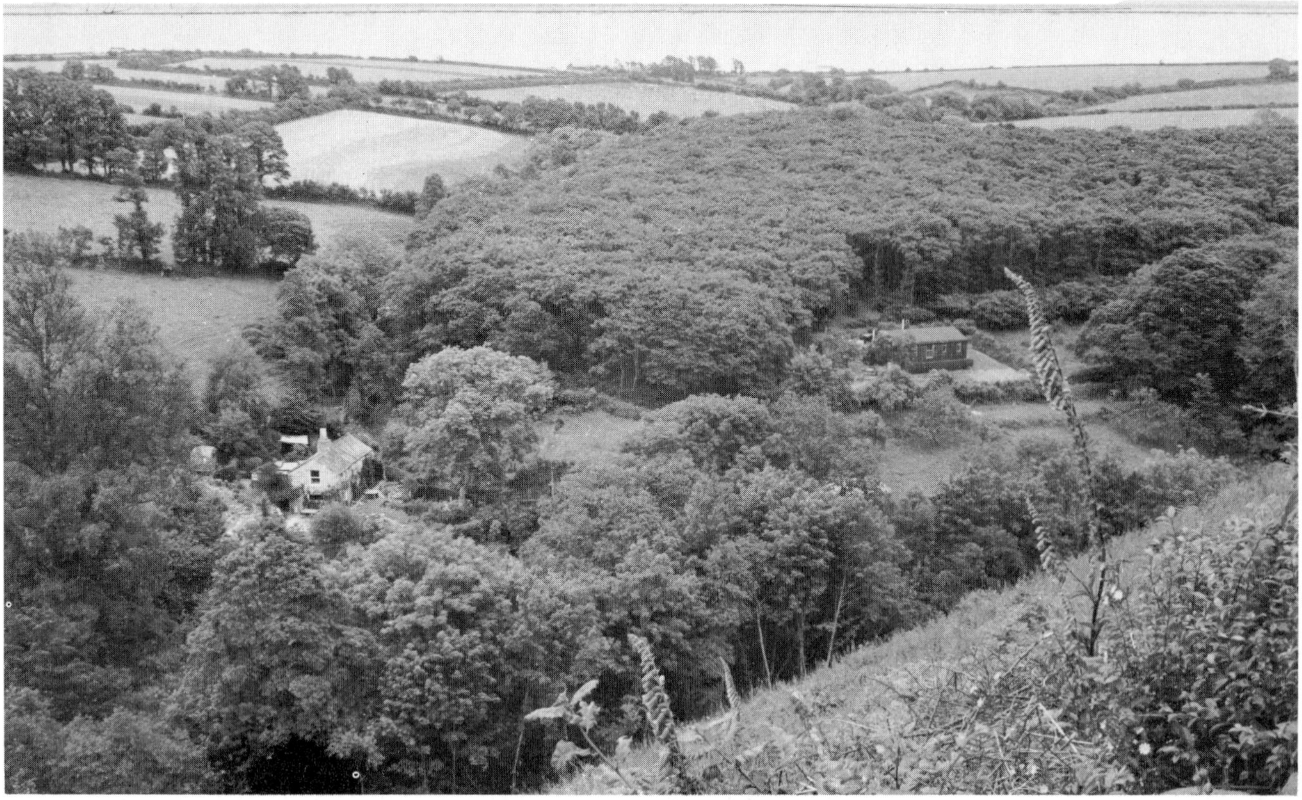

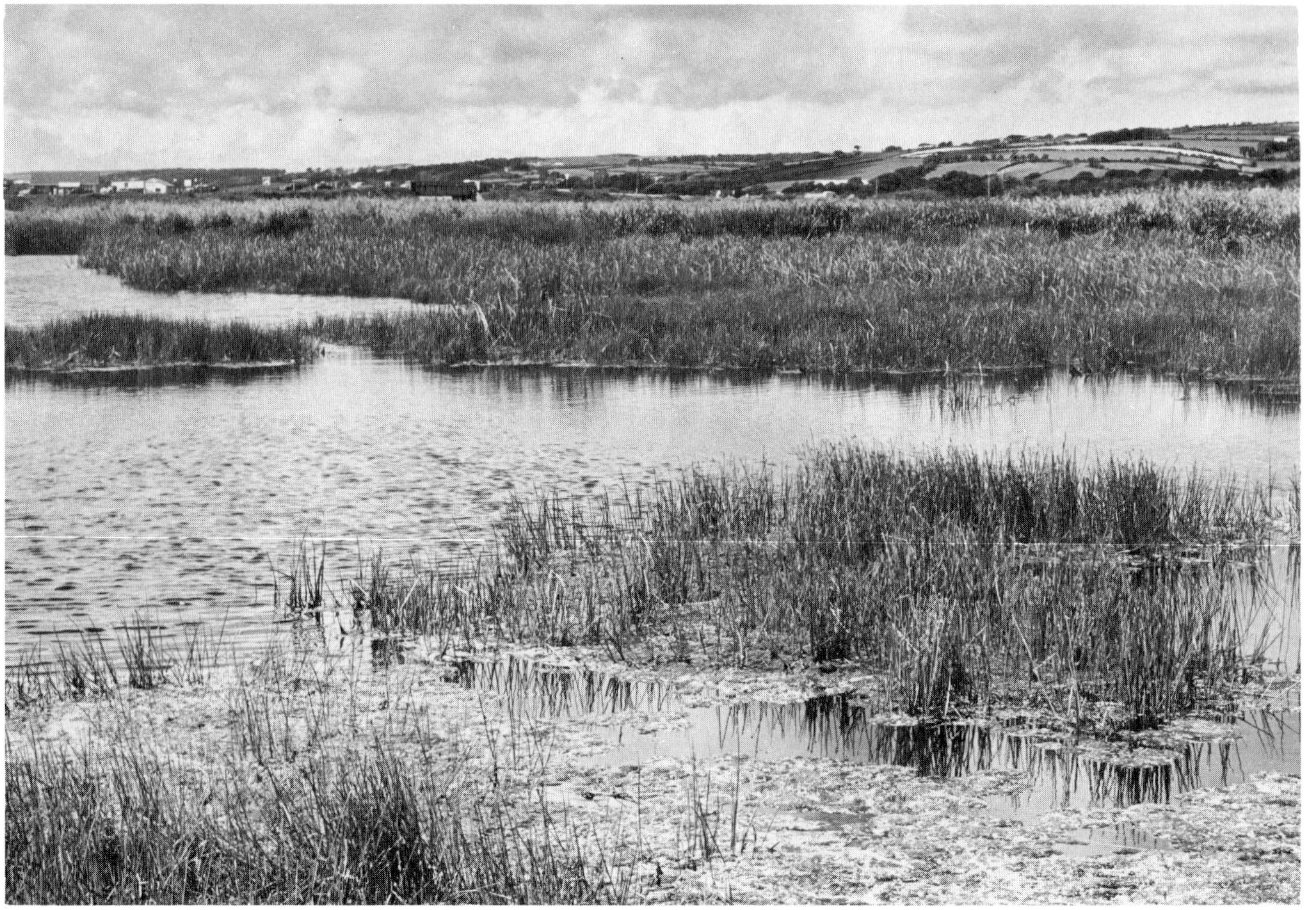

OPPOSITE ABOVE: Hawkes Wood, near Wadebridge, the first Trust
Reserve — note trees growing in the traditional hedges in the background;
BELOW: Marazion Marsh. (JB & SB) ABOVE and LEFT: Erosion through
'people pollution' at Lands End, (NT) and RIGHT: the flooded Kit Hill
Quarry, near Callington. (Jill Long)

ABOVE: Bulrushes on Red Moor Reserve; (NCC) LEFT: Trelissick House
set within its extensive parkland, (NT) and RIGHT: The now extinct Large
Blue butterfly once abundant in certain coastal valleys. It is the emblem of
the Cornwall Naturalists' Trust. (D.C. Hockin)
OPPOSITE: Maer Cliff on the culm coast. (H. Tempest/NT)

Appendix I: Reserves of the Cornwall Naturalists' Trust

I. TRUST FREEHOLD

Hawkes Wood (SW 986709) near Wadebridge; a nine acre wood on sloping ground with an old quarry, which supports a luxuriant growth of ferns, and a stream. The wood is mainly coppiced oak with sycamore, ash and a typical ground flora; an excellent area for woodland birds.

Hayman Reserve, Park Hoskyn (SW 748497) near St Agnes; a wet and steep sided valley with sessile oak, chestnut and ash. There is a marshy area, a small quarry and a disused mine-shaft to produce an exceptional variety of habitats within a small compass (4.6 acres). There is varied bird and insect life; resident mammals include the water shrew.

Kemyel Crease (SW 460244) near Lamorna; an unusual area, five acres in extent, of cliff woodland formerly used for flower growing. It is divided into numerous small plots, and the flora includes both woodland and coastal species growing in close proximity; there are cypressus and other exotics planted for shelter together with indigenous vegetation.

Red Moor (SX 073622) near Lostwithiel; essentially a heathland area (50 acres) with some woodland, pools and a lake created by past mining excavations. The result is a wide range of habitats which support an unusual variety of plants, birds and other animals especially reptiles, amphibians and spiders. The reserve was purchased with help from the World Wildlife Fund and the Cornwall Bird-Watching and Preservation Society as a memorial to Lt Colonel W.E. Almond, Mrs E.W.G. Campbell and Dr D.L. Johnson.

Ventongimps Moor (SW 781512) near Perranporth; an area of 20 acres which constitutes an exceptionally good example of wet heath and lowland bog with an excellent flora, which includes Dorset heath, and associated insect life. There are problems of dessication and consequent gorse encroachment.

2. RESERVES HELD BY LEASE, LICENCE OR AGREEMENT WITH LANDOWNERS

Drift Reservoir (SW 436288) near Penzance; approximately 90 acres by lease from South West Water Authority. This covers margins of the reservoir within the permieter fence and certain rights over the reservoir itself. The bird life is of considerable interest.

Fal-Ruan Estuary: Two sections of the estuary (Trelonk, SW 890405, and Ardevora, SW 881406) near Ruan Lanihorne, leased from the National Trust and covering approximately three miles (250 acres) of muddy foreshore and salt-marsh partly on china clay silt; not only is there interesting vegetation but the reserve forms a valuable feeding ground for wading birds and wildfowl.

Kynance Cliffs and Lower Predannack Downs (SW 688132) on the Lizard, 215 acres leased from the National Trust. The area is mainly open heathland and cliff slopes on serpentine rock with Cornish heath and other examples of the unqiue Lizard flora.

Lower Predannack Cliffs and Mullion Island (SW 660163) on the Lizard, 41 acres leased from the National Trust. The area is mainly lightly grazed heathy cliff slopes on hornblende schists with typical Lizard flora. The reserve includes three isolated headlands (Pedn Crifton, Men-te-heal and Laden Ceyn) the last of which is serpentine. Mullion Island supports a large colony of breeding sea birds, notably kittiwakes.

Luckett Reserve (SW 392728) Stoke Climsland, nine acres leased from the Duchy of Cornwall. The reserve is a natural woodland area within a complex of reafforestation, bounded on one side by a stream. It is of exceptional interest in respect of both plants and insects, particularly Lepidoptera.

Pelyn Woods (SX 088588), Lanlivery near Lostwithiel, 102 acres held by agreement with the owners — Mr and Mrs Nicholas Kendall. The main block of woodland consists largely of beech with oak, ash, a good ground flora and a stream fringed by ferns. A smaller secondary block includes marshy ground and an area of sessible oakwood underlain by granite. There are badgers, other mammals and considerable entomological interest.

Pelyne (SX 178598), Lanreath, 9.2 acres held by agreement with the Dartington Hall Trust. The reserve consists of a wooded valley, a small marsh and varied habitats typical of south Cornwall.

Pendarves Wood (SW 641377), south of Camborne, 44½ acres held under a management agreement with the Forestry Commission. The area is deciduous woodland, partly replanted with conifers, and includes a lake of some five acres which supports varied and interesting flora and fauna. There is a nature trail and development for educational use.

Peter's Wood (SX 113910), near Boscastle, 25 acres leased from the National Trust. The reserve is part of the impressive wooded Valency valley and has been little disturbed by man except for past coppicing. It is a varied mixed deciduous wood with both sessile and common oak as well as other trees and an abundant ground flora. There is an unusually large mollusc population probably because of the underlying volcanic rock which provides calcium for their shells.

Porthcothan Valley (SW 868713), St Eval, sixteen acres held by agreement with the owner, Mr T.O. Darke. This reserve consists of a sheltered valley with a stream, a small meadow, a disused quarry and some woodland — habitats very typical of valleys near the north Cornwall coast.

Tamar Estuary, approximately six miles (c.1000 acres) of tidal foreshore and mud-flats alongside the Tamar river, both north and south of Landulph (SX 432630), leased from the Duchy of Cornwall. The mud-flats, between high and low water marks, are used as a feeding ground by a large variety of wading birds and wildfowl often in impressive numbers; interesting salt-marsh vegetation. The reserve includes Kingsmill Lake.

Tremelling Reserve (SW 350341) near St Erth, nine acres held by agreement with the owner, Miss Loveday. The reserve consists of marshy ground, willow carr and some standing water in the Hayle river valley; the flora and fauna is typical of such habitats.

Treworra Farm (SX 154866) near Davidstow. By agreement with the owner, Mr T. Burgess, the whole of this 180 acre typical moorland farm, is treated as a sanctuary for the plentiful wildlife. It includes bog, wet heath, two upland streams, a rookery and an ancient sunken lane where there is an abundant and varied hedgerow flora.

(Except where normal rights of way exist, persons wishing to visit these reserves are advised that before doing so, they should consult either the administrative or conservation officers of the Trust or the district representative. Honorary wardens are responsible to the Trust for the day by day management of all reserves.)

3. OTHER RESERVES

There are protected areas in Cornwall other than Trust reserves. Foremost among these is *The Lizard National Nature Reserve,* established with statutory authority by Nature Conservancy Council and covering an area of over 1,000 acres. It includes several different kinds of Lizard heathland and is divided between Predannack Cliff (SW 660168), Goonhilly Downs, Traboe Downs, Ponsongath, Crousa Common and Main Dale (SW 786204).

The Cornwall Bird-Watching and Conservation Society owns *Walmsley Sanctuary* (SW 996747) a 42 acre section of the Amble marshes which form an arm of Camel estuary; many species of duck, goose and wader are attracted by a wide variety of aquatic habitats. The society also owns *Trethias Island* (SW 855739), a large off-shore stack about two miles south of Trevose Head. It holds the shooting rights, which are not exercised, over *Hayle Estuary* (SW 550380) and part of *Restronguet Creek* (SW 810385) and leases from the South West Water Authority part of *Stithians Reservoir* (SW 714385) probably the most important stretch of inland water in Cornwall for waders and wintering wildfowl — certain other reservoirs, including the *Tamar Lakes* (SS 295112) have been declared as Bird Sanctuaries. In co-operation with the Trust and the Cornwall Bird-Watching and Preservation Society, the Water Authority is planning (1981) to give reserve status to the northern *Loveny River* (SX 184740) section of Colliford Reservoir, approximately 250 acres of mainly shallow water and moorland verge.

The National Trust owns or protects by covenant an overall area of 24,082 acres in Cornwall. While this land is not specifically nature reserve, it is protected from development and is a major element in the protection of nature in Cornwall. Details of properties can be found in the appropriate National Trust publications, readily available in National Trust shops.

Appendix 2: Sites and Areas of Ecological Interest

The list which follows includes all Sites of Special Scientific Interest (SSSIs) in Cornwall, other than reserves shown in Appendix 1, with a representative selection of the Trust's Conservation Sites and other areas of interest. These places are not necessarilly open to the public. Many are on private land, and the owner's permission should be sought before paying a visit. The coast path passes through a majority of the coastal areas, and access to beaches and open moorland is uncontrolled. Wherever you are the Country Code should be honoured; wild birds and other animals should not be disturbed; growing plants should not be uprooted, nor should they be picked in excessive numbers — some rare plants are totally protected; dogs should be kept on leads in places where domestic animals are present.

   Grid references mostly refer to a central or dominant point within the site but bounding limits have usually been given where long stretches of coast have been considered as a unit.

Battery Rocks, Mounts Bay (SW 490305) near Penzance; seashore site with rocks and stones showing varying degrees of exposure; dense seaweeds and abundant fauna, especially barnacles.

Bedruthan Steps and Park Head (SW 843710 — SW 846689) north of Mawgan Porth; spectacular rock scenery with cliff vegetation showing the effects of extreme exposure — partly owned by National Trust.

Boconnoc Park and Woods (SX 144603) near Lostwithiel; an area of deciduous woodland and ancient deer park; in addition to the normal wildlife of such habitats, the site supports an outstandingly rich epiphytic lichen flora.

Bodmin Moor (Brown Willy SX 159798) the highest and most extensive of Cornwall's granite moorlands with open moor and rough grazing land, clitter slopes, rocky tors, bogs, streams, standing water and wooded verges; the moor is cut into two major sections by the A30 trunk road. North of the road is the highest and most rugged area with Brown Willy (1,375 feet) and Rough Tor, Crowdy marsh and reservoir of outstanding botanical and ornithological interest, major bogs and river sources. South of the road, more land has been reclaimed and enclosed but this section of the moor includes Dozmary Pool (SX 195745) with interesting aquatic and marginal plants, the Loveny (Colliford reservoir) and Lamelgate valleys, Kilmar Tor (SX 253748), the marshy Twelve Men's Moor, the fine upland stream of Withey Brook, Halvana plantation (SX 210790) with Forestry Commission trail, the mixed woodlands at North Hill where the moor drops to the Lynher river, and the granite with topaz quarry (SX 259724) below Stowe's Hill where the Cheesewring stands.

Boscastle to Widemouth Bay (SX 093915 — SX 195019), a spectacular stretch of culm coast of considerable geological interest; zigzag folding of the strata on Millook cliff, stunted oak forest of the Dizzard, Dizzard Point, High Cliff, the Strangles (see below) and Beeny cliffs with large grey seal breeding colony; cliff and cliff-top vegetation and bird life are of particular interest — partly owned by National Trust.

Breney Common (SX 055613) close to Red Moor reserve and including similar habitats; primarily dry heath and with wet areas and willow carr providing cover and nesting places for a wide range of birds; formerly worked extensively for alluvial tin.

Bude Canal Marshes (SS 208060), to the south of the town and including both canal and River Neet; as well as the two very different waterways there are reed beds, marshy ground, meadows and trees of varying height to provide a wide range of habitats; includes District Council nature reserve.

Camel's Cove with The Straythe (SW 931388) near Nare Head; a sheltered south coast cove with interesting cliff vegetation; excellent example of a sheltered beach with interrupted reefs and some partly exposed areas with mussel beds as well as plentiful seaweeds and the southern barnacle — partly owned by National Trust.

Camel Estuary (SW 990725 — SW 920800) below Wadebridge; the one major break in the north coast cliffs; a major tidal estuary with salt-marsh, mud-flats and dunes, notably those of recent origin at Rock; ancient raised beach with fossil deposits at Trebetherick Point (SW 926780); subsidiary creeks include Amble river and marshes. The whole estuary is of outstanding importance for birds, particularly wildfowl and waders, there being many good observation points; this interest continues above Wadebridge — short sections owned by National Trust.

Cape Cornwall to Clodgy Point (SW 353313 — SW 508412), the granite cliffs of the north coast of Penwith which includes Botallack (old mine buildings and headland), Pendeen Watch, Rosemergy and Bosigran cliffs and Zenor Head; spectacular scenery with typical cliff vegetation and breeding sea birds; grey seals breed and regularly haul out on the Carracks (SW 467409) — partly owned by National Trust.

Cardinham Woods (SX 105680) near Bodmin; an extensive area of Forestry Commission plantations with some hardwood areas; nature trail.

Carnkief Pond (SW 787520) near Perranporth; pool with adjacent boggy woodland; abundant ferns and typical bog vegetation.

Carrine Common (SW 795431) near Truro; a comparatively small area of dry heathland with a good growth of Dorset heath — this species normally thrives in much wetter conditions.

Chyenhal Moor (SW 448279) near Penzance; a wet moor on the southern slopes of the Penwith granite; various habitats including several small ponds; acquatic and bog vegetation are of interest.

Clicker Tor Quarry (SX 285614) near Liskeard; the quarry reveals exposures of an igneous rock much used for Stone Age artefacts.

Cligga Head and Pen-a-Gader (SW 737530) north of St Agnes Head; 300 foot cliffs where there are exposures of both granite and killas rocks; good maritime heath at the southern end of the site which has been much prospected and mined for various minerals.

Coombe Valley (SS 225115) north of Bude; Forestry Commission plantations with some deciduous woodland; nature trail managed by the Trust.

Cotehele (SX 4233685) near Calstock; National Trust house and gardens with nature trail in woods beside River Tamar.

Delabole Quarry (SX 075840) near Camelford; huge, deep slate quarry which has been worked for centuries; fossil brachiopods (Delabole 'butterflies') are to be found in certain slates.

Dinas Head and Constantine Bay (SW 858750) near St Merryn; headland with colourful cliff vegetation and nesting birds; relatively stable dune area with interesting flora and insect life.

Dodman Point (SX 002397) near Gorran Haven; a striking and little disturbed south coast headland with varied plant and animal life; thick cliff-top scrub provides shelter for migrant birds — National Trust property.

Draynes Wood (SX 222685) near St Neot; a damp oak and ash wood on steep slopes beside the River Fowey and around Golitha Falls; dripping rock faces and abundant plant life — the best remaining example of an old wood on the edge of Bodmin Moor.

Duckpool (SS 200115) north of Bude; an exposed shore with some shelter provided by rocky reefs; abundant mussels and exceptional colonies of sedentary polychaete worm, *Sabellaria alveolata,* with sand-tubes grouped in masses.

Dunmere and Pencarrow Woods (SX 043687) near Bodmin; an extensive area of plantations (Forestry Commission and private) with some natural woodland running down to the River Camel; red deer.

Eglarooze Cliffs (SX 349539) near Portwrinkle; a stretch of south coast cliffs with distinctive vegetation including a rare broomrape.

Fal Estuary (SW 850405), a complex system which embraces the Rivers Fal, Tresillian and Truro with Malpas estuary (SW 835430) and includes the Trust's Fal-Ruan reserves; extensive mud-flats and salt-marsh of considerable botanical and ornithological interest; large areas of deciduous woodland beside the rivers — some National Trust property on banks.

Feock Beach (SW 826381) four miles south of Truro; a good example of a sheltered beach with rocks, stones and gravelly sand; dense belts of seaweeds.

Fowey River (SX 170810 — SX 125510) one of the finest rivers in the county; rising on the high moor, passing through Lamelgate valley, Draynes Wood, Lostwithiel and the Lerryn woods as well as part of Lanhydrock estate and woodland, the river finally enters the sea at Fowey; the banks of the river support a cross-section of the wildlife of Cornwall.

Gannel Estuary and Crantock Beach (SW 805608) near Newquay; small estuary with some salt-marsh and a minor dune system; attracts a number of interesting birds — National Trust property.

Gerrans Bay and Nare Head (SX 915370) near Veryan; a major south coast headland which demonstrates many of the geological features of the county and includes the longest continuous stretch of raised beach in Cornwall; there is considerable botanical interest — largely National Trust property.

Godrevy Head to St Agnes (SW 582422 — SW 710518), the coast on either side of Portreath and north of Camborne (Camborne North Cliffs); widely contrasting vegetation types from dune flora to cliff grassland and maritime heath; striking range of flowering plants and several interesting bird communities; grey seals breed —largely National Trust property.

Goss Moor (SW 950600) lies beside the A30 trunk road, south-east of St Columb Major; lowland heath and pasture land with pools, some willow carr, old mine buildings and waste heaps; the area is of considerable botanical and ornithological interest.

Greenscombe Valley Woods (SX 395730) near Stoke Climsland; Duchy of Cornwall woodland with conifer and hardwood plantations in combination; good populations of woodland birds and interesting butterflies; the Trust's Luckett Reserve is part of this complex.

Hayle Estuary and St Ives Bay (SW 553375) between Hayle and St Ives; extensive mud-flats form an important feeding ground for a great variety of wading birds; though much damaged by human pressure the Hayle-Gwithian dune complex, around the bay, still supports a highly characteristic flora and fauna, particularly on the Upton Towans.

Helford River (SW 707268) south of Falmouth; estuary and creeks with tidal mud-flats and wooded banks; Merthen Wood (SW 730263) being an outstanding example of a western sessile oakwood; interesting birds and marine animals; seal sanctuary at Gweek (SW 707268) — some National Trust property.

Hensbarrow Downs (SW 995575), the St Austell granite with china clay workings throughout the area; some isolated patches of untouched heathland; natural recolonisation of some older tips and modern efforts at reclamation and replanting can be seen at many places.

Herodsfoot Woods (SX 205604) south-west of Liskeard; Forestry Commission plantations and deciduous woodland; trees include red oak; red deer and other animals.

Hessenford Woods (SX 305573) north of Seaton and athwart the Seaton river; mixed woodland area, now largely replanted but supporting a varied fauna.

Isles of Scilly, an extension of the granite mass of Cornwall; separated by shallow seas, the islands support an unusual flora and are of outstanding interest to ornithologists; breeding area of grey seals and home of Scilly shrew; seashore life of both exposed and sheltered beaches; access restricted to certain outlying islands including Annet which is a bird sanctuary.

Kelsey Head and Holywell Bay (SW 770600) between Newquay and Perranporth; wide range of coastal habitats including cliff-top grassland and sand dunes; maritime plants plentiful; breeding auks and other sea birds, particularly on off-shore stacks — largely owned by National Trust.

Kilminorth Wood (SX 245543) near West Looe and beside tidal reach of the river; mixed deciduous woodland with large population of nesting birds.

Land's End (SW 343251) has been greatly damaged by human trampling but the exposed granite cliffs support a fine display of 'splash zone' plants including colourful lichens; Sennen Cove (SW 352263) immediately to the north is a sandy beach below granite cliffs with good maritime vegetation; further north again, between Port Nanven and Cape Cornwall (SW 318350) the rocky shore shows well the effects of extreme exposure with dense barnacles but few mussels — some National Trust property.

Lanlydrock (SX 085636) near Bodmin; a fine park and mixed woodland areas with varied and plentiful flora and fauna including nesting birds; nature trail — National Trust property.

The Lizard (SW 700200) peninsula south of Helston; National Nature Reserve and the two Trust reserves cover only a small part of the Lizard which is of extreme interest throughout; extensive heaths and downs (including Kynance cliffs and cove, SW 684132, Lizard Downs, SW 695140, Goonhilly, SW 735195, and Traboe, SW 745208, Downs) mostly on serpentine rock support an remarkable flora including Cornish heath and many rare plants with Lusitanian associations; Ruan Pool (SW 697158), Croft Pascoe (SW 731198) and Hayle Kimbro (SW 695170) pools bring aquatic conditions to the area; the geology of the peninsula is unique in Britain, particularly along the superb coast line parts of which are owned by the National Trust.

Loe Pool (SW 647250) near Porthleven; the largest freshwater lake in Cornwall cut off from the sea by a shingle bar; proximity to the sea and the surrounding marshes and woodland mean a wide variety of habitats close together; interesting flora and bird life, notably wintering wildfowl; the pool is owned by the National Trust and is protected from disturbance by fishing, boating and swimming.

Loggans Moor (SW 577390) near Hayle; a good example of a pasture meadow with typical flora — such meadows are still quite plentiful in many parts of Cornwall.

Looe with Porthnadler Bay (SX 240510, the position of Hore Stone) is a relatively sheltered rocky beach with numerous pools; abundant seaweeds and typical fauna of this type of shore with many golden star tunicates or sea-squirts.

Lynher Estuary (SX 372568) enters the Tamar between Saltash and Torpoint and includes Erth Island; varied estuarine habitats with mud-flats, salt-marsh, freshwater of the river and fringing woods (notably Sheviock Wood, SX 370562) lead to an unusually diverse bird life.

Marazion Marsh (SW 515318) near Penzance; the remains of a much larger marshy areas which, however, still supports the largest bed of *Phragmites* reeds in Cornwall; some open water and the adjacent seashore makes the area unusually attractive to a wide variety of birds.

Marsland Mouth and Valley (SS 213175) on Devon border; Marsland Mouth shows a great variety of slope, soil and aspect, and supports a considerable diversity of maritime vegetation; the valley is occupied by deciduous woodland for a distance of nearly three miles and is mostly owned by the Royal Society for Nature Conservation.

Minions (SX 265715) near Liskeard; an exceptional concentration of old mine buildings, on northern flank of Caradon Hill, surrounded by land in different stages of recolonisation by natural vegetation.

Mulberry Down Quarry (SX 019658) west of Bodmin; geological site previously used for open-cast mining.

Nance Wood (SW 665450) near Portreath; dwarf sessile oaks with varied ground flora which includes a rare spurge.

Newlyn Downs and Penhallow Moor (SW 835550) east of Perranporth; a good example of natural recolonisation of old mine tips and land covered by metalliferous waste.

Newquay Beaches (SW 798625) exposed beaches to the north and west of the town with shell sand and rocky reefs supporting dense barnacles and large mussel beds.

Par Sands and Marsh (SX 085533) near St Austell; sandy beach on seaward side of marsh which has been partly reclaimed but still attracts numerous birds.

Penhale Sands (SW 771572) or Perranporth dunes; the highest sand-dunes in Britain, three miles long and more than a mile wide; though considerable damage has been done by human disturbance, parts of the system support a highly characteristic flora, the result of high lime content of the sand.

Penwith Moors (from Trencrom Hill, SW 518363, to Carn Kenidjack, SW 385335) include Rosewall (SW 488391), Trendrine (SW 479387) and Mulfra (SW 450345) hills, Bosullow (SW 418345) and Woon Gumpus (SW 400340) commons with Bostraze marsh (SW 393320) and other wet heath and boggy areas; though Watch Croft (SW 421358) the highest point is only 828 feet, this moorland is generally more rugged than most of Bodmin Moor with lichen-covered boulders, gorse and heathers; plentiful signs of former mining as well as of occupation by Bronze Age and Stone Age Man.

Place Cove (SW 847323) in St Mawes inlet; a sheltered rocky shore showing good zonation of seaweeds and interesting fauna including varied mollusc populations and polychaete bristle-worms.

Polperro West Cliffs (SX 202505) near Polperro; well vegetated south coast cliffs with considerable botanical interest; maritime plants and several rare species — mostly owned by National Trust.

Porkellis Moor (SW 690323) between Stithians and Helston; a lowland heath area with acid soils and a marked diversity of habitats derived from former mining activity; flora and fauna, particularly dragonflies, are of interest.

Porth Reservoir (SW 870620) near Newquay; adjacent woodland brings many birds to the area in addition to the wintering wildfowl and wading birds.

Porthgwarra to Pordenack Point (SW 371217 — SW 345241) the stretch of coast immediately south-east of Land's End with Gwennap Head and Carn Les Boel; an extensive, spectacular and extremely exposed stretch of granite cliffs with maritime heathland and a fine display of coastal flowers; breeding sea birds on the cliffs and good viewing points for passage migrants; Porthgwarra valley provides good shelter for small birds — partly convenanated to National Trust.

Porthleven Cliffs (SW 622258); the geological interest centres upon the 'Giant's Stone' an erratic boulder of highly polished gneiss of a type found nowhere else in Britain.

Porthscatho to Greeb Point (SW 878344) at the end of Gerrans Bay; low, sheltered south coast cliffs with interesting maritime flora.

Praa Sands (SW 580280) west of Helston; a small area of sand dunes, which are not frequent on the south coast, with a good variety of wild flowers.

Rame Head (SX 418483) to the east of Whitsand Bay and south of the Tamar estuary; a prominent headland with a good variety of wild flowers; a spectacular path follows the cliff edge.

Restronguet Creek (SW 810380) between Truro and Falmouth; a tidal river which enters the Carrick Roads; extensive mud-flats much used as a feeding ground for wading birds.

Rosenannon Bog (SW 959662) about three miles north-east of St Columb Major; small bog supporting a typical flora; there are several bogs and marshes in this area which lies to the south of St Breock Downs (SW 965680) now largely reclaimed.

St Agnes Beacon Pits (SW 705510) below St Agnes Beacon; the the best exposures in Cornwall of Pliocene gravels.

St Erth Sand Pits (SW 537351) at St Erth, south of Hayle; old sand pits with deposits of Pliocene fossils — site now rather overgrown.

St Ives Island (SW 520413) St Ives headland; an excellent place for watching sea birds on migration, particularly during the autumn.

St John's Lake (SX 424540) Torpoint; an extensive area of tidal mud-flats in the Tamar-Lynher estuary complex; supports large populations of wintering wildfowl.

Silverwell Moor (SW 748482) near St Agnes; an area of wet moorland where there is a good growth of Dorset heath.

Steeple Point to Marsland Mouth (SS 198116 — SS 210173) the cliffs both north and south of Morwenstow, the site being continuous with the Duckpool and Marsland Valley sites; a long stretch of rocky coast on the culm with sheer cliffs, examples of vertical stratification and a wide range of coastal habitats; varied flora and insect life as well as breeding birds — partly owned by National Trust.

Stithians Reservoir (SW 715364) near Stithians; attracts larger flocks of wintering wildfowl than any other stretch of freshwater in Cornwall; seasonal mud at water's edge provides ideal conditions for wading birds on passage.

Stourscombe Quarry (SX 344839) near Launceston; a geological site in the cherts and slate beds with fossil exposures.

Strangles (SX 130955) near Crackington Haven; cliff slopes of botanical interest form part of the Boscastle to Widemouth Bay site; sandy and rocky beaches below are excellent examples of exposed rocky shore with dense barnacles and mussel beds.

Swannacott Woods (SS 250980) near Week St Mary; large area of Forestry Commission plantation in a part of Cornwall where small deciduous woods are still plentiful; red deer.

Talland Bay (SX 223515) between Looe and Polperro; coastal grassland with good flora and patches of scrub running down to the beach; numerous small birds; a typical sheltered south coast beach and bay with plentiful sea urchins.

Tamar Estuary (SX 450535 — SX 425681) from mouth of the river to Cotehele Quay; the site includes Tamar Estuary reserve and adjoins St John's Lake and Lynher estuary sites; together they form the largest area of mud-flats in south-west Britain; of particular interest in respect of wading birds (including wintering avocets) and wildfowl as well as for salt-marsh vegetation.

Tamar Lakes (SS 290115) near Kilkhampton and athwart the Devon border — old Tamar Lake with new Tamar Lake (reservoir) immediately above; bird sanctuary since 1949 and Regional Wildfowl Refuge; while the particular interest is in wintering wildfowl, many other birds are present as well as aquatic and verge plants and insect life; large freshwater mussels.

Tintagel Cliffs (SX 044862 — SX 090913) from south of Trebarwith Strand to Boscastle; a spectacular and almost unbroken stretch of cliffs with fine display of coastal plants; sea birds, notably on Lye Rock (SX 064899) where there is the largest puffinry in Cornwall, and the Trevalga cliffs (SX 078909) with auks breeding on off-shore stacks — partly owned by National Trust.

Treen Cliffs (SW 393224) on south coast of Penwith; includes Treryn Dinas cliff castle and rugged granite headland; cliff-top scrub provides shelter for small birds — National Trust property.

Tregargus Quarry (SW 749541) west of St Austell; a deep quarry with exposures of china stone within the St Austell granite.

Trelissick (SW 837396) above Feock and near Truro; gardens, park, pasture woodland and deciduous woodland beside the River Fal — National Trust property.

Trenowth Wood (SW 931510) near Grampound Road; a western sessile oak wood, partly replanted, on the edge of the St Austell granite and cut by the upper Fal river.

Trevone Beach (SW 887760) at Trevone, west of Padstow; good example of a partly exposed rocky shore with sand and broken rocky reefs covered with limpets, barnacles and mussels; many rock pools.

Wrasford and Eastcott Moors (SS 260145) on the culm north of Kilkampton and source area of the River Tamar; the most northerly moors in Cornwall, now mainly pasture with *Juncus* rushes and some patches of mixed heath; formerly a breeding area of Montagu's harrier.

Appendix 3: Officers of the Cornwall Naturalists' Trust

President

Dr F.A. Turk	1962-1967
Rennie M. Bere, CMG	1967-1970
Colonel W.E. Almond	1970-1973
Dr C.J.F. Coombs	1973-1976
Dr K.S. Hocking	1976-1980
Dr Colin G. Butler, OBE, FRS	1980-

Vice-President

Dr F.H.N. Smith	1962-1969
Dr C.J.F. Coombs	1969-1973
Mrs Enid Campbell	1973-1974
A.D. Ellory	1975-1978
Dr Colin G. Butler, OBE, FRS	1978-1980
Dr F.H.N. Smith	1980-
Dr D.L. Johnson	1964-1974
Mrs C.M. Johnson	1974-1979
Bryan Wilson	1979-

Chairman of Council

Colonel W.E. Almond	1962-1970
Dr D.L. Johnson	1970-1975
Dr K.S. Hocking	1975-1976
Philip Blamey	1976-

Vice-Chairman of Council

Dr Gillian Matthews	1962-1969
Dr D.L. Johnson	1969-1970
Dr K.S. Hocking	1970-1975
Philip Blamey	1975-1976
Dr F.H.N. Smith	1976-1980
Mrs C.M. Johnson	1980-

Hon Treasurer

M.H. Bizley	1962-1963
C.R. Bartlett	1963-1966
A.D. Ellory	1966-1969
M.F. Francis	1969-1973
M.C. Hawken	1973-1981
J.T. Williams	1981-

Hon Secretary

J.K. Williams	1962-
J.H. Prideaux (acting)	1974

Hon Membership Secretary

Mrs M. Tregenza	1969-1971
Mrs B. Stanley	1971-1976
Mrs C.M. Johnson	1976-

Hon Sales Officer

Dr D.L. Johnson	1964-1974
Mrs C.M. Johnson	1974-1979
Bryan Wilson	1979-

Editor of the News Letter

Colonel W.E. Almond	1962-1974
T.O. Darke	1974-1981
W.F.H. Ansell	1981-1982

Hon Legal Adviser

P.A.S. Pool	1962-1979
Richard P. Jones	1979-

Representative on County Trusts' Committee of R.S.N.C.

Colonel W.E. Almond	1962-1971
Dr K.S. Hocking	1971-

Administrative Officer

W.F.H. Ansell	1974-

Conservation Officer

Caroline Rigby	1979-

Bibliography

Almond, W.E.. *Distribution of Bumble-Bees in Cornwall and the Isles of Scilly* (Institute of Cornish Studies, 1975).

Andrews, A.W. *Poems of West Penrith* (Saundry, Penzance, 1957).

Andrews, A.W. and Pyatt, E.C. *Cornwall* (Climbers' Club, 1950).

Arnold, E.N. and Burton, J.A. *A Field Guide to the Reptiles and Amphibians* (Collins, 1978).

Arnold, H.M. *Provisional Atlas of the Mammals of the British Isles* (N.E.R.C., 1978).

Axford, E.C. *Bodmin Moor* (David and Charles, 1978).

Barr, John *Derelict Britain* (Penguin, 1969).

Barrett, John and Yonge, C.M. *Pocket Guide to the Seashore* (Collins, 1958).

Barton, R.M. *A History of the Cornish China Clay Industry* (Bradford Barton, 1966).

Barton, R.M. *An Introduction to the Geology of Cornwall* (Bradford Barton, 1964).

Bere, Rennie *Wildlife in Cornwall* (Bradford Barton, 1970).

Blamey, Philip and Marjorie *Marjorie Blamey's Flowers of the Countryside* (Collins, 1980).

Brewster, Carolyn *Bodmin Moor: A Synoptic Study on a Moorland Area* (Institute of Cornish Studies, 1975).

Brian, M.V. *Ants* (Collins, 1977).

Bristowe, W.S. *The World of Spiders* (Collins, 1958).

Burrows, R.S. *The Naturalist in Devon and Cornwall* (David and Charles, 1971).

Burrows, R.S. and Butts, R.B. *Wildlife of the Land's End Peninsula* (Cornwall County Council, 1980).

Burton, John *The Oxford Book of Insects* (Oxford, 1968).

Calder, Harry *County Structure Plan* (Cornwall County Council, 1979).

Campbell, A.C. *Guide to the Seashore and Shallow Seas of Britain and Europe* (Hamlyn, 1976).

Campbell, Bruce and Watson, Donald *The Oxford Book of Birds* (Oxford, 1964).

Chapman, V.J. *Coastal Vegetation* (Pergamon, 1964).

Clegg, John *Freshwater Life of the British Isles* (Frederick Warne, 1965).

Clegg, John *The Observer's Book of Pond Life* (Frederick Warne, 1967).

Colyer, Charles N. and Hammond, Cyril *Flies of the British Isles* (Frederick Warne, 1951).

Coombs, Franklin *The Crows* (Batsford, 1978).

Corbet, G.B. and Southern, H.N. (eds.) *The Handbook of British Mammals* (Blackwell, 1977).

Darke, T.O. *The Cornish Chough* (Bradford Barton, 1971).

Davey, F. Hamilton *Flora of Cornwall* (Chegwidden, 1909 — EP Publishing Ltd., 1978).

Dickinson, C.L. *British Seaweeds* (Eyre & Spottiswood, 1963).

Earl, Bryan *Cornish Mining* (Bradford Barton, 1968).

Findlay, W.P.K. *Wayside and Woodland Fungi* (Frederick Warne, 1967).

Fitter, Richard *Dictionary of British Natural History* (Penguin, 1967).

Fitter, Richard, Fitter, Alastair and Blamey, Marjorie *The Wildflowers of Britain and Northern Europe* (Collins, 1974).

Ford, R.L.E. *The Oberver's Book of Larger Moths* (Frederick Warne, 1963).

Free, John B. and Butler, Colin G. *Bumblebees* (Collins, 1959).

Gill, Crispin *The Isles of Scilly* (David and Charles, 1975).

Hamilton, W.R., Woolley, A.R. and Bishop, A.C. *Guide to Minerals, Rocks and Fossils* (Hamlyn, 1973).

Hammond, Cyril W. *The Dragonflies of Great Britain and Ireland* (Curwen Press, 1977).

Hardy, Alister *The Open Sea* (Collins, 1956).

Harris, Stephen *The Secret Life of the Harvest Mouse* (Hamlyn, 1979).

Heath, John *Provisional Atlas of the Insects of the British Isles — Butterflies* (N.E.R.C., 1970).

Hepburn, Ian *Flowers of the Coast* (Collins, 1952).

Hewer, H.R. *British Seals* (Collins, 1974).

Higgins, L.G. and Riley, N.D. *A Field Guide to the Butterflies of Britain and Europe* (Collins, 1970).

Hudson, K. *The History of English China Clays* (David and Charles, 1971).

Imms, A.D. *Insect Natural History* (Collins, 1947).

Johns, C.A. *A Week at the Lizard* (S.P.C.K., 1848).

Johns, C.A. — ed. by Blakelock, R.A. *Flowers of the Field* (Routledge and Kegan Paul, 1949).

Kershaw, K.A. and Alvin, K.L. *The Observer's Book of Lichens* (Frederick Warne, 1963).

Lange, Morton and Hora, F. Bayard *Guide to Mushrooms and Toadstools* (Collins, 1965).

Laws, Peter *A Guide to the National Trust in Devon and Cornwall* (David and Charles, 1978).

Larousse Encyclopedia of Animal Life (Hamlyn, 1967).

Lenton, E.J., Chanin, P.R.F. and Jefferies, D.J. *Otter Survey of England, 1977-79* (N.E.R.C., 1981).

Lewis, J.R. *The Ecology of Rocky Shores* (English Universities Press, 1964).

Lockett, G.H. and Millidge, A.F. *British Spiders* (Vols. 1 &2, The Ray Society, 1951 and 1953).

Locket, G.H., Millidge, A.F. and Merrett, P. *British Spiders* (Vol. 3, The Ray Society, 1974).

Lyneborg, Leif translated by Vevers, G. and Reade, Winwood *Mammals in Colour* (Blandford, 1971).

Mabey, Richard *The Common Ground* (Hutchinson, 1980).

Macan, T.T. and Worthington, E.B. *Life in Lakes and Rivers* (Collins, 1951).

Margetts, L.J. and David, R.W. *A Review of the Cornish Flora* (Institute of Cornish Studies, 1981).

Martin, W. Keble *The Concise British Flora in Colour* (Ebury Press, 1965).

Matthews, Gillian and Parks, Peter *Seashore Life* (Puffin, 1965).

McClintock, David and Fitter, R.S.R. *Pocket Guide to the Wildflowers* (Collins, 1963).

Mellanby, H. *Animal Life in Fresh Water* (Methuen, 1948).

Mercer, Ian *Nature Guide to the West Country* (Usborne, 1981).

Mitchell, Alan *A Field Guide to the Trees of Britain and Northern Europe* (Collins, 1974).

National Trust in Cornwall, The (Skyshots, St Ives, 1980).

Neal, Ernest *Woodland Ecology* (Heinemann, 1953).

Newman, H. and Mansell, E. *The Complete British Butterflies in Colour* (Ebury Press, 1968).

Nichols, D., Cooke, J. and Whitely, D. *The Oxford Book of Invertebrates* (Oxford, 1971).

Paton, Jean A. *A Bryophyte Flora of Cornwall* (British Bryological Society, 1969).

Paton, Jean A. *Wild Flowers in Cornwall* (Bradford Barton, 1968).

Pearsall, W.H. *Mountains and Moorland* (Collins, 1950).

Penhallurick, R.D. *Birds of the Cornish Coast* (Bradford Barton, 1969).

Penhallurick, R.D. *The Birds of Cornwall and the Isles of Scilly* (Headland, 1978).

Peterson, R.T., Mountfort, G. and Hollom, P.A.D. *A Field Guide to the Birds of Britain and Europe* (Collins, 1958).

Pollard, E., Hooper, M. and Moore, N.W. *Hedges* (Collins, 1974).

Pyatt, E.C. *Coastal Paths of the South West* (David and Charles, 1971).

Rackham, Oliver *Trees and Woodland in the British Landscape* (Dent, 1977).

Ragge, David R. *Grasshoppers, Crickets and Cockroaches of the British Isles* (Frederick Warne, 1965).

Reade, Winwood and Hosking, Eric *Nesting Birds* (Blandford, 1967).

Rose, Francis an Stokoe, W.J. *The Observer's Book of Ferns* (Frederick Warne, 1965).

Ryves, B.H. *Bird Life in Cornwall* (Collins, 1948).

Sankey, J.H.P. and Savory, T.H. *British Harvestmen* (Linnean Society, 1974).

Sheldon, J.C. and Bradshawe, A.D. *Hydraulic Seeding Techniques for Unstable Sand Slopes* (Journal of Applied Ecology, 1977).

Shorter, A.H., Ravenhill, W.L.D. and Gregory, K.J. *South West England* (Nelson, 1969).

South, Richard *The Butterflies of the British Isles* (Frederick Warne, 1941 ed.).

South, Richard *The Moths of the British Isles* (Frederick Warne, 1961 ed.).

Stamp, L. Dudley *Britain's Structure and Scenery* (Collins, 1946).

Stamp, L. Dudley *The Common Lands of England and Wales* (Collins, 1963).

Stamp, L. Dudley *Nature Conservation in Britain* (Collins, 1969).

Steers, J.A. *The Sea Coast* (Collins, 1953).

Thurston, E. and Vigurs, C.E. *A Supplement of Davey's Flora* (Royal Institute of Cornwall, 1922).

Tregarthen, J.C. *Wildlife at the Land's End* (Murray, 1904).

Turk, Frank *Distribution Patterns of the Mammalian Fauna of Cornwall* (Institute of Cornish Studies, 1973).

Turk, Stella *Sea Shore Life In Cornwall* (Bradford Barton, 1970).

Various Authors *Handbooks for the Identification of British Insects* (Royal Entomoligical Society of London).

Watson, Alan *Butterflies* (Kingfisher Books, Ward Lock, 1981).

Whalley, Paul *Butterfly Watching* (Severn House, 1980).

Woolf, Charles *Archaeology in Cornwall* (Bradford Barton, 1970).

Yonge, C.M. *The Sea Shore* (Collins, 1949).

Newsletters and Bulletins of Cornwall Naturalists' Trust.

Annual Reports of Cornwall Bird Watching and Preservation Society.

Cornish Biological Records (Institute of Cornish Studies).

Index

General Index

Fauna Index

Flora Index

N.B. Appendix material is not reflected in these Indices.

Subscribers

Presentation Copies

1 HRH Prince Charles, Prince of Wales
2 The Cornwall Naturalists' Trust
3 Cornwall County Council
4 The Royal Society for Nature Conservation
5 Nature Conservancy Council
6 Cornwall County Library
7 The National Trust

8 Philip & Marjorie Blamey
9 Rennie Bere
10 Clive & Carolyn Birch
11 Franklin Coombs
12 Pat Paton
13 Gordon Bottomley
14 Charles Woolf
15 Colin Butler
16 Kenneth Williams
17 Christopher Cadbury
18 Stephen Reeves
19 Graham Henderson
20 Institute of Cornish Studies
21 Polperro C. Primary School
22 W.D. Broadbent
23 Miss I. Martin
24 Ken & Ethel West
25 Mike Hasshil
26 St Columb Minor Jnr School
27 J.B. & S. Bottomley
28 Maree & Rennie Bere
31
32 P. Truscott
33 A.D. Ellory
34 Freda E.H. Bramley
35 Mrs Vanessa Beeman
36 Peter Chard
37 Mrs Shelagh Garrard
38 Malcolm Read
39 Victoria & Albert Museum
40 B.S. Kiek
41 Dr & Mrs K.S. Hocking
42
43 Julie Hockin
44 Mrs R.J. Dutton
45 Dr P.J.G. Butler
46 Burness & Cyril Bunn
47 F.M.P. Pearce
48 R.L.D. Pearce
49 F.B.P. Pearce
50 Constance Swain
51 D.C. Trehane
52 L.J. Margetts
53 R.H. Wills
54 Brian Wilson
55 Mrs Teresa Pearce

56 Raymond Dennis
57 B.E.M. Garratt
58 Elizabeth Tregenza
59
60 Pat Griffiths
61 A. Hepworth
62 Miss M.K. Burgess
63 Geoffrey & Pamela Sprunt
64 P.M. Holligan
66
67 J.M. Horrell
68 Philip Currah
69 C.R. Henderson
70 W. Stuart Best
71 Mrs C.H. Underwood
72 Miss R. Cole
73 C.D. Preston
74 Miss Hilary Westgarth
75 B.J. Leach
76 Miss D.J. Andrew
77 Dr T.C. Marks
78 David Pugh
79 Dr Ernest G. Neal
80 Mrs U.L.E. Jefferis
81 F.P. Cozens
83
84 Mrs Joan Ham
85 Prof S.G. Kiriakoff FRES
86 Miss H. Wadham
87 Mr & Mrs G. Connett
88 Mrs K. Warren
89 F.W.C. Merritt
90 Mrs Kegowing
91 Mrs D.G. Squires
92 Mrs Katherine Rowe
93 E.W.M. Magor
94 Nigel J. Crocker MBOU
95 J.R. Hart
96
97 Mrs K.M. Thomas
98
99 Anne Pearce
100 Adrian & Sharon Stark
101 Martin Sebastian Ward
102 M.J.H. Tonking
103 T.C.H. Retallack
104 Sir Peter Hunt

105 Leslie Joy Hyde
106 Mrs M. Hale
107 M.V. Pulley
108 Bernard Moore
109 Dick Twinney
110 Paul Sokoloff MSc, MIBiol, FRES
111 B.L.J. Byerley FRES
112 Mary E.T. Watts
113 Dr Bruce Ing
114 Mrs S.M. Pinsent
115 I.M. Wilkinson
116 David Smith
117 Barbara E. Selley
118 Adrian M.B. Butterworth
119 Joan E. Judge
120 Mrs C. Lane
121 Miss K. Haly Richards
122 Ms L. Baylis
123 J.C. Hill
124 I. Dracup
125 John R. Hawell
126 Dr K.A. Monk
127 R.S.R. Fitter
128 Mr J.W. Reece
129 E. Nicholl
130 E.A. George
131 John S. Wallis
132 Steven Lance Hunt
133 Mrs M.A. Adams
134 J.S. Ryland
135 P. Maddison
136 Dr A.R. Brain
137 M. Ellis MD
138 Mrs J. Eaton
139 Roger & Sheila Peters
140 G.W. Burgin
141 Mrs Valerie Pettifer
142 Miss M.C. Carter
143 Edward Bolger
144 Mrs R. Parfitt
145 Mr & Mrs R.M. Tetley
146 Eric W. Harding
147 Ian Taylor
148 A.J. King
149 K. Badcock
150 H. Margaret A.H. Shirley
151 F. Fincher
152 Cecily Ashcroft

153 Mr & Mrs G.E. Hood
154 Jean & Laurie Tester
155 Veronica Brendon
156 Miss B.M. Sturdy
157 Dr William J. Carter
158 Sir David & Lady Willcocks
159 Susannah Mulcock
160 Callington CP School
161 Elizabeth J. March
162 Janet A. Warren
163
164 Bude Junior School
165 Peggy Visick
166
167 Bruce Burley
168 David M. Barber
169 C.M.K. Stanton
170 Margaret Edith Rochford
171 B. Blowers
172 J.E. Libby
173 Claire Shaddick
174 Anne-Marie Ellis
175 John Blowey
176 Felicity M. Lang
177 William Thomas Wain
178 R.P. Treneer
179 G.C. Treneer
180 Stuart V. Robinson
181 G.D. Trebilcock
182 Susan M. Nundy
183 The Rev A. Mapplebeck
184 D.S. Burrows
185 Jill Long
186 A.E. Price
187 Alan Horton
188 Mrs J. Pearce
189 Reginald J.C. Williams
190
191 Church Town Farm, Lanlivery
192 Sarah Jackson
193 G. Jackson
194 Mrs M. Strong
195 Penwith Sixth Form College
196 Nick Brown
197 Peter & Mary Coleman

198 Lindsay Henry
199 Richard P.H. Coleman
200 Mrs M. Gordon
201 Sandy Hill CP School
202 R.N. Hesketh
203 Adrian Rutter
204 Mrs Kathleen Rutter
205 Jacobstow CP School
206 D. Hartley-Russell
207 Mrs Sylvia D. Johns
208 R.H. Trenoweth
209 L.F. Blacknell
210 Keith Alexander
211 E.V. Bowden
212 Miss I. Rosemary
     Frost
213 Mrs Anne E. Freer
214 Bernice Briggs
215 Elizabeth Higgs
216 Richard Arthur
     Tribbeck
217 A.C. Leslie
218 Jean & John Dawkins
219 Mrs H. Woodford
220 Mrs B. Eardley
221 F.W. Shepherd
222 Mrs D.G. Ede
223 Dr A. Nelson-Smith
224 Dr P.E. King
225 T.D.R. Hockaday
226 Miss A.M. Richards
227
228 M.E. Solomon
229 G.W. Harper
230 E.C. Pelham-Clinton
231 Mrs J.M. Lynch
232 Dr G.N. Foster
233 H. John Dickinson
234 C.C. Wilcock
235 Anne Hedley
236 Mrs P.C. Stafford
237 R.J. Bridport
238 A.J. Bird
239 William Cresswell
240 Brian Hiley
241 Miss Jean E.A.
     Hemens
242 V. Knight
243 Christopher Bailey
244 R.D. Butson
245 L.H. Roberson
246 John Hough
247 Mrs B.A. McConville
248 Philip & Marjorie
     Blamey
249 Mrs I.M. Catchpole
250 Ivan Rabey
251 Peter Charles Searle
252 A. Sumner
253 Mrs W.E.P. Miller
254 M.M. Linforth
255 Mrs Anne Gilbert
256 Mrs D.O. Cox
257 Noreen E. Sherlock
258 Carol Ohlenschlager
259 Mrs J.M. Butt

260 Charles Philip Wheater
261 Paul Attlee
262 Martin Lazell
263 J.H. Adams
264 H.A. Deal
265 Muriel Thirsk
266 Dr Robert P. Higgs
267 R.J. Richards
268 Mrs E. Brooke
     Houghton
269 Peter F. Marfadyen
270 Mrs A.M. Bythway
271 A.M. Downing
272
273 Dr P.R. Hatherley
274 Mark & Clare Kitchen
275 Captain E.T. Bolitho
276 A.R. Bolitho
277
278 Mrs M. Giles
279 Mrs P.M. Pelham
280 M.J. West
281
282 R.M.P. Lyne
283 Caroline Ruth Wherry
284 Miss R.E. Lees
285 Ralph Hall Calvert
286 M.J. Clemens
287 Miss K.M. Rushton
288 H. Fine
289 Mrs Lucinda Cattran
290 Anne Irons
291 Robin Blamey
292 Timothy Blamey
293 Amanda Verrin
294 Mrs Madge Davis
295 Jill Palmer
296 Mrs H.M. Malin
297 Miss B. Stephens
298 J.B. Winlove-Smith
299 B. Gray
300 Mr & Mrs J.I.S.
     Henderson
301
302 J.M. Milner FLE
303 Dr R.G. Balf
304 K.A. Hendry MBE
305 S.G. Hedge
306 Miss J. Port
307 D.I. Lockie
308 Mrs Pat Snow
309 John P. Wreford
310 Kathleen Mary Kedgh
311 M.W. Mann
312 Denise Ramsay
313 Mrs J.M. Barkla
314 Mrs L.A. Cadbury
315 T.O. Darke
316 W.S. Barnard
317 L.J. Riley
318 C.D. Wallis
319 G.A. Pearce
320
321 Miss M. Stephenson
322 Miss T.K. Sykes
323 C.S. North

324 Miss Karen
     Greenwood
325 Eckhard Moller
326 W.R. Davies
327 Ross Walker
328 Mr & Mrs V. Hicks
329 Jackie Raymer
330 Kevan Rudling
331 I.G. McCulloch
332 J.T. Williams
333
334 Mrs Ann Spencer
335
336 Mrs F. Nankivell
337 Prof George Trease
338 Dr S. Gardner
339 M.C. Livesley
340 Mrs Rosemary
     Coward
341 Ruth Walker
342 J. Ellis
343 Miss D. Waterson
344 Prof J. Green
345 G.R. Else
346 Arthur Bryant
347 Miss R.V. Diprose
348
349 Miss M. Bunt
350 Mrs S.W. Shepley
351 John Archer
352 G.C. Steele
353 Jose Beer
354 Mrs E.C. Bassett
355 T.J. Bassett
356 Mr & Mrs S. Brain
357 Miss M.B.W. Davies
358 Bryan W. Moore
359 Deborah Lewis
360 Mrs Eileen Curtis
361
362 Mrs. R. M. Dingle
363 Miss M. Nelson
364
366 Restormel Branch CNT
367 Spencer G. Howlett
368 Barbara Featherstone
369 Mrs V.L.F. Skerratt
370 Miss A. Raymond
371 M. Wilkinson
372 Dr D. Middleton
373 Mrs P.H. Brown
374 Sir Frederick Bishop
375
376 David B. Owens
377 Andrew C.V. Hawke
378 Mr & Mrs P.W.G.
     Coombs
379 Ray Woodbine
380 J.M. Moore
381 St Nicholas CE
     School, Torpoint
382 R.K. Merrifield
383 St Catherines C of E
     School, Launceston
384 Mr & Mrs D. Brown
385 Jean I. Slateford

386 Mrs Elizabeth Beare
387 Mrs Mary Jones
388 K.M. Hallam
389 R. Smaldon
390
     Blisland CP School,
     Bodmin
391
392 R. Purser
393 A. King
394 A. Body
395
397 A.J.R. Rule
398 St Stephen CP School,
     St Austell
399 Mrs Margaret White
400 Mrs D.M. Fox
401 Francis Hewlett
402 Susan Dow
403 T.C. & M. Burnard
404 Brian Thomas
405 Constance Doreen
     Alison
406 Lynda B. Pembleton
407 Georgina L. Firth
408 Miss A. Jeffreys
409 Mrs B. Wilding
410 Leonard Ennis
411 R.M. Radcliffe
412 S.C. Hutchings
413 Mr & Mrs A.F.
     Reynolds
414 Miss Ann Reynolds
415 Peter J. Blackburn
416 J.O. Meaby
417 Miss L.W. Turpitt
418 Miss M. Perry
419 J.M. Randall
420 Mr & Mrs K. Bendell
421 O.C.A. Cornelius
422 H.J. Ingrey
423 J.A. Sage
424 Sidney H. Hill
425 Miss D.A. Phillips
426 G.C. Jackson
427 Dr G.P.G. Rowe
428 Mr & Mrs C. Robinson
429 Doris Lambert
430
431 Mrs Mary Page
432 Mrs Betty Gaite
433 Elizabeth M. Carne
434 William Pennycook
435 K.A. Stevens
436 H.M. Wilsdon
437 Mrs S. Cusworth
438 Miss J. Greenham
439 Dr J. Spurrier
440 M. Walpole
441 John P. Wreford
442 Michael Lyth
443 A.P. Barnes
444
446 Dr John Burston
447 Mrs M.K. Frost
448 Mrs M.A. Smith
449 B.T. Craven

450 John Fanshawe
451 Jane Davis
452 Douglas A. Grose
453 B.W. Bousfield
454
455 Miss J.F. Hawes
456 Bernard Skinner
457 Mrs E.M. White
458 Miss Zoe K. Potter
459 Winifred A. Currie
460 Mrs M. Gwyneth Norris
461 John Heath
462 A.E. Smith
463 The Very Reverend Frank Curtis
464 Diana & Ben Taylor
465 W.E. Holloway
466
467 Mrs D.L. Brookman
468 Cyril West
469 H.W. Mackworth-Praed
470 Mrs I. Winn
471 Mrs O.M. Hepworth
472 R.W. David
473 Rev A.C. Canner
474 T.M. Hotten
475 Mrs G. Hopkins
476 S.A. Boase
477 R. Torry
478 E. Mottershead
479
481 Barry A. Jones
482 Alice C. Bizley
483 P.D. Hewett
484 Margaret J. Rowe
485 Miss Lucy K. Overbury
486 Maree Bere
487 B. Prichard
488 Prof W.G.V. Balchin
489 Mrs H. Davoll
490 Mrs M.J. Allen
491 Roger E. Warrier
492 Jonathan Fenton
493 Dorothy J. Herlihy
494 Mrs Letsey Wood
495 D. Griffin
496 Mrs M.R. Pybus
497 Patricia Foster
498 Dr N.R. Maslen
499 Phyllis M. Cook
500 Johnny Geeves
501 Stephen Harris
502 Mr & Mrs J. Goddard
503 David & Anita Robinson
504
505 Miss M.H. Whitehead
506 Leslie James Wellard
507 Veronica Thres
508 Mrs M. Griffin
509 Miss M.P. Anderson
510 Miss E. Finlay

511 R.A. French
512 T.D. Walker
513 Phillippa S. Robinson
514 Anthony Holden
515 Barbara J. Squire
516 Mr & Mrs M.R. Brooks
517 Mrs A. Elston
518 Hilary Fowler
519 N.S. Roseby
520 John E. Medlock
521 Dr & Mrs B. Taylor
522 Alan Christopher Gange
523 David W.R. Thackray
524 Margaret Martin
525 Miss H. Naish
526 P.J. Roderick
527 Mrs Nancy Linley
528 Miss V. Strout
529 Mr & Mrs J. Foot
530 J. Hodge
531 J.E. Jagger
532 R. Fairclough
533 W.H. Batten
534 M.G. Watkins
535 Audrey Lord
536 Win Randall
537 Christine Appleyard
538 J.A. Keene
539 H.E. Cox
540
541 D.C. Hockin
542 Mrs K. Winter
543 Mr & Mrs H. Davy Thomas
544 Mr & Mrs D.W. Bodley
545 Mrs D. del Nicholls
546 L.N. Baxter FRES
547 Pamela Sekula
548 Linda Mary Philp
549 Mrs N. Bowyer-Smith
550 J.W. Moore
551 Margaret Budd
552 E.F. Nicholls
554 D.A.J. Little
555 Arnold John Nicholls
556 P.G. Lamerton
557 Mrs Edna Sexton
558 R.E. Wake
559 S. Greenhour
560 A.E.L. Greenhour
561 M.J.L. Greenhour
562 Patrick E. Coleman
563 David G. Sharpe
564
565 Mrs A. Keers
566 P. Hambly
567
568 Jeremy Beare
569 Reginald A. Long
570 Elizabeth Deverell
571 Christine J. Dodd
572 Dr Mary Gillham
573 Derek R. Williams

574 S. Brennan
575 Richard Hurst Poynton
576 Rex Bray
577 Lady Reed
578 Edward Browning
579 Mr & Mrs M.E. Greenhalgh
580 Mrs A.E. Barfett
581 P.T. Barfett
582 Barbara Wilson
583 Arnold Royston Morgan Bradley
584 Paul Toynton
585 Mrs Evelyn L. Almond
586 Mr & Mrs Howell
587 Dr Keith Hammond
588 Dr Michael J. Palmer
589 D.L. Sparrow
590 A.R. & A.J. Codrington
591 Institute of Terrestrial Ecology
592 Mrs N. Bersey
593 Miss P. Keats
594 Hugh Wylie
595 J.M. Clatworthy
596 P.B. Lindsey
597 A.D. Tomlin
598 Eileen W. Beall
599 H.P.K. Robinson
600 Miss D.F. Metcalfe
601 Mr & Mrs H.J. Stanlake
602 G.P. Gill
603 P.K. Bray
604
605 Stewart Balke
606 Michele Robson
607 Michael Parsons
608 A.C. Cutter
609 Miss M. Maclerie
610 Mrs Rosanne Barber
611 Marian Padgett
612 Dr John V. Turner
613 Mrs B.E. Benney
614 B.J. Leach
615 Miss Eve Northey
616 Mardi Tempest
617 V. Cowan
618 Miss I. Cowley
619 Sidney A.A. Painter
620 Olga Warwick
621 Mrs Helen Derrington
622 Margaret Nunn
623 Christopher John Nunn
624 Sarah Jane Constance
625 Richard James Nunn
626 J. Eric Hunt FICS
627 Vanessa J. Hunt
628
629 W.I. Dickinson
639 Dr M.H. Shere
640 Mrs K.N. Quarterman
641 R.W. Gill

642 Mrs L.H. Tongue
643 Dr A.G. Sangster
644 Mrs B.D. Kerr
645 Mrs E. Cooper
646 Mrs J.E. Hicks
647 J.C. Wolters
648 Mrs M. Tomlinson
649 M.F. Phelan
650 Mrs R.E. Carey
651 Michael Allaby
652 Mrs D. Murley
653 L.A. Cram
654 Charles Woolf
655 F. Brian J. Coombes
656 H.R.H. Lance
657 Mrs M. Parslow
658 E.D. Vowles
659 Muriel F. Barnett
660 Valerie Clark
661 Maree Bere
662 P.J. Wonnacott
663 Michael William Tyler
664 James V. Newark
665 The Rev H.C.S. Fowler
666 T.J.S. Pinfield
667 Mrs R.C. Redding
668 D.A. Polgrean
669 L.I. Hamilton
670 T.C.E. Wells
671 D.R.J. Chahe
672 Sqn Ldr A.R. Marshall
673 B. Stratford
674 Dr B.N. & Mrs D.M. Eagles
675 Patrick Rawlinson
676 S.P. Nicholls
677 Mr & Mrs W.F. Haines
678 Stephen Watkins
679 R.J. Walkey
680 B.E. Keech
681 F.W. Shepherd
682 Mrs P. Hersant
683 Laura Andrew
684 Mrs Mary Adeane
685 Mrs D. Childs
686 G.A. Garceau
687 Mr & Mrs E. Emery
688 W.D. Broadbent
689 R.P. Cutler
690 R.S.C. Copeland
691 N.P. Cummins
692 G.D. Penhallorick
693
694 J.A.R. Arnold
695 B. Holdaway
696 Cherrith K. Pearce
697 Mrs N.J. Osborne
698 Ladock School, Truro
699 J.C. Dakin
700 R.L. Philbey
701 Jean Wingfield
702 Dr & Mrs F.H.N. Smith

703 J. Jerome
704 Rosemary E. Broughton
705 Mrs R.P. Weeks
706 Mr & Mrs R.S.E. Edmunds
707 A.E. Purkiss
708 Vivien Charlton
709 S.R. Harper & S.J. Hebidge
710 N.F. Hormbrey
711 Dr Janet Hormbrey
712 Mrs M. Evans
713 Wendy Wottan
714 Mrs P.G. Turpin
715 P.J. Terry BSC
716 Mrs Cynthia Garner
717 D. Steel
718 Mrs Phyllis E. Blackall
719 Winifred Lewin
720 Dr V. Challinor Davies
721 Mrs B.M. Brown
722 D. McClintock
723 David House
724 John G. Pickwell
725 E. Dearing
726 Mr & Mrs Carswell
727 Prof R.J. Berry
728 Jane G. King
729 Stella M. Turk
730 Miss J.M. Gregory
731
732 Dr B.P. Thurlow
733 Godfrey Worraker
734 David N. Robinson
735 Mr & Mrs R.A. Snoxall
736 Dr Eric Duffey
737 Mrs Carolyn Keep
738 John Beswick
739 Col V.I. Fisher
740 Mary Waller
741 John H. Drew
742 J.N. Frankland
743 Dr Ann Bailey
744 K.J. Burrow
745 Mrs Elizabeth Mico
746 Mrs S. Palmer
747 P.M. Jackson
748 J.W. Eaton
749
751 Mrs E. Willoughby
752 M. Godwin
753 Mrs J.W.A. Posnett
754 Mrs C.K. Jones-Parry
755 Peter & Nan. Hicks
756 D.H. Ridler
757 Dr A.P. Lewis
758
759 Nancy Dawson
760 Mr & Mrs David G. Spear
761 Matthew Hyde
762 Mrs Sonia C. Holland
763 Mrs Jean M. Benfield

764 B.J. Williams
765 Mrs C.A. Hogg
766 Mrs G. Stringer
767 Truro Cathedral School
768 Rosemary J. Lomas
769 Miss V.M. Hughes
770 Muriel & Ron Joslin
771 Miss Connie Saunders
772 A.G. Cavell
773 W.J.S. Hosking
774
775 D.C.N. Smith
776 L.H. Roberson
777 Mrs Dennis Burford
778 Jane Parkinson-Maddocks
779 J.K. Williams
780 Owen Willoughby
781 Mrs Ethel Willoughby
782 Peter Lee
783 Dr Lawrence A. Storer MB ChB FRAS
784 Mrs F.C. Morgan
785 P.M. Turner
786 Mrs K.R. Hodge
787 J. Spurway
788 Dr R.M. Bell
789 J.W. Muhn
790 P.W. Stone
791 Capt R.B.N. Hicks
792 Rosemary Dick
793 Evelyn E. Atkins
794 A.R.L. Rouse
795 Noel Stuart
796 A.A. Langford
797 Molly Morgan
798 Mrs C.M. Johnson
799 J.G.S. Turner
800 P.J. Dwyer
801 Mrs K.J.C. Milligan
802 General Library, British Museum (Natural History)
803 Mr & Mrs S.A. Rowland
804 Miss I.H.S. Shaw
805 Mrs S.M. White
806 Mrs E.D. White
807 Evangeline Rouncefield Berriman
808
810 G.C. Thorley
811 J.D.C. Muirhead
812 F.P. Wright
813 Mrs T.M. Stephens
814 Miss P.R. Osburn
815 Miss H.B. Trant MBE
816 Miss R.E.B. Trant
817 A.D.M. Cox
818 B.J. Prichard
819 Miss Mabel Lobb
820 Geoff Milburn
821
823 Pat Macmillan

824 Mrs H.R. Graham-Vivian
825 Mrs A.C. Ham
826 Miss R.M. Phillips
827 Iris Patricia Sugg
828 T. Wilson
829
830 Z.D. Hosking
831 Sqd Ldr G.T. Witherick
832 D.H. Draper
833 Mrs J.C. Gynn
834 C. Robert Skinner
835 Mrs Irene Weston
836 R.T.M. Wareham
837 Richard Pearce
838 T.O. Darke
839 Mrs Nora M. Coleman
840 Frank W. Leigh
841 J.R. Jeffery
842 Mrs M.L. Bettesworth
843 Marion Williams
844 Mrs W.J. Facey
845 Rev B.J. Freeth
846 P.J. Sangster
847 Mr & Mrs J. Small
848 John A. Vaughan
849 Mrs N.B. Groggon
850 Paul Chanin
851 Mrs Margaret Ruth
852 Mary Martin
853 Mary Avent
854 Mrs M.C. Tout
855 Janet & John Weekes
856 Elizabeth D. Earl
857 Frances Judy Lemos
858 Mrs M.V. Strong
859 Mr & Mrs M. Rosewell
860 David B. Irving
861 John M.R. Irving
862 Mrs Kay Orriss
863 Dr C. Wiblin
864 Philip Michael Harris
865
866 Dr & Mrs P.H. Stevens
867 A.E.C. Aston
868 Mrs Noreen L. Barron
869 Jill & Tony Yates
870 Dr D. Jackson
871 Mrs B.T. Walker
872 Bosivgo Primary School, Truro
873
874 Mrs J. de Raaf
875 Chalewater CP School, Truro
876 Avon County Library
877 Mrs J. Rimmer
878 M.P. Lumley
879 J.O.J. Stevens
880 E.W.F. Tomlin
881 Church Town Farm Lanlivery
882 W.F.H. Ansell
883 John Klima
884 W.A.R. Wolfenden

885 Barbara Wills
886 Mrs G.F. Caunt
887 Mrs Leveen J. Hill
888 Enid Davey
889 Alyson Joy Bradley
890 Brian R. Tuck
891 Catriona B. Beale
892 J.D. Wedd
893 R.S. Brenchley
894 Ralph Hal Calvert
895 Kenneth Kendall
896 Dr Sally Corbet
897
898 Miss M.C. Botterell
899 G. Roberts
900 Dilys Philips
901
907 Dr C.J.F. Coombs
908 Susan Truscott
909 Margaret Ingles
910 Mr & Mrs Clark
911 Iris Bennett
912
913 M.C. Toms
914 Iris Warne
915 Jill T. Trembeth
916 Patsi Bettenson
917 Daphne White
918 Mrs M. Martindale
919 Angela J. Adams-Rice
920 Rev Ian S. Holdworth
921 Mr & Mrs C.H. Blamey
922 Gilbert John Bryant
923 G. Gynn
924 Carolyn Channing
925 Miss R.J. Gallop
926 Mr & Mrs D.V. Jessup
927 P.N. Bowden
928 E. & J. McKenzie
929 Mr & Mrs S.C. Robinson
930 Mr & Mrs J.C. Lobb
931 Mr & Mrs A.T. Beswetherick
932 Miss M.K. Hawkins
933 J.P. Treglown MPS ARPS
934 Miss Jennifer Wade
935 Dr Brian John Prout
936 Dr Claire Harries
937 Mrs J.F.M. Smedley
938 M.E. Edwards
939 Alison Adburgham
940 Steve Jackson
941 W.C.F. Knight
942 Teresa Budge
943
944 The National Trust
945 Mrs Barbara Atyeo
946 Miss D.F. Metcalfe
947 Mrs J.P. Gill
948 Miss Muriel A. Briggs
949 Mrs Audrey Lucas
950 L. Forsyth

951 Mrs Dorothy Houston
952 B. June Bentley
953 C.L. Robbins
954 J.L. & Mrs J.F.M. Smedley
955 Nature Conservancy Council
956 Valerie Burt Marston
957 A.G. Holman
958 J. Howard Curnow
959 J. Hendy
960 R.G. & P.K. James
961 Christine North
962 Barry G. Tattersall
963 Cheryll Pitt
964 Mrs J.E. Brown
965 L.W. Turpitt
966 Keith Plummer
967 Dave Snell
968 Judy V. Cope
969 Dr Susan M.W. Vaidya
970 Dorrit B. Smith
971 Mrs G.H. Willcocks
972 Mrs Sheila Skeet
973 N.J. & S.M. Chambers
974 R.P.S. & J.S. Maclean
975 T.C. & D.M. Tydeman
976 Robin Hanbury-Tenison
977
978 Miss C. Hilary Bates
979 N.N. Beach
980 Miss W.M. Povey
981 Miss R.M. Jerram
982 Mr & Mrs K.M. Leach
983 The Royal Institute of Cornwall
984 Sherry S. Lee
985
986 J.R. Hunt
987 Michael Grist
988 David Leslie Thomas
989 C.K. Tallack
990 Miss M.E. Hutchins
991 J.M. Moore
992 Philippa Van Langendonck
993 Miss Sheila M. Keily
994 N.W. Duncan
995 R.A. Hodgkin
996 Joan Moffatt
997 C.F. Tomaszewski
998 W.T. Manning
999 Roy F. Arnold
1000 Col J.P. Carne
1001 Stephen P. Beech
1002 T.T. Bartlett
1003 W. Stuart Best
1004 Kathleen M. Oliver
1005 R.J.I. Moon
1006 John E. Blackmore
1007 Mrs Eileen Bomford
1008 David Negus
1009 Julia Drage

1010 Alma Hathway
1011 R.J. Beswetherick
1012 John Henry Woods
1013
1014 Mrs Rachael Dingle
1015 Geoff Milburn
1016 Miss Enid Harvey
1017 Mary Donovan
1018
1019 Mrs M.J. Brown
1020 D.A. Harding
1021 M.D. Ames
1022 Dr M.H. Martin
1023 Dr L.C. Frost
1024 R.J. Salmon
1025 Mrs Lois Blades
1026 Edward Griffiths
1027 Heather M.A. Bishop
1028 Richard David Hawken
1029 Mrs J.K. Higman
1030 Roger Butts
1031 Stella Plasom
1032 Mrs Frances Smith
1033 W.B. Champion
1034 Brian Sheen
1035 Mrs Moller
1036 K.M. Thomas
1037 Mrs V.K. Lavelle
1038 Kay Jones
1039 R. Mahoney
1040 Mrs E.M. Roberts
1041 M. & P. Collins
1042
1043 R.A. Middleton
1044 Ronald A. Weston
1045
1046 G. Creighton Balfour
1047 Mrs C.I. Reynolds
1048 Sidney H. Hill
1049 E.D. Earl
1050 Mrs Joy Etherington
1051 T.S. Reid
1052 Doreen Saunders
1053 Mrs P. Alderson
1054 C.S. Robbins
1055 T.A. Lordin
1056 J. Martin
1057
1058 B.D. Stephens
1059 Mrs G.S. Lyall
1060
1061 C.D. Vanstone
1062 Mrs C. Desborough
1063 Ann Salsworthy
1064 B.S. Hursey
1065 Mrs Wheatley
1066 Adrian Langdon
1067 Mrs A.S. Moulder
1068 W.H. Woolcock
1069 Mrs B. Bassett
1070 Mrs A. Warne
1071 Miss M. Garland
1072 Mrs F.J. Hooper
1073 Mrs Heather Cox
1074 Mrs J. Holman

1075 C.D. Corbett
1076 C.F. Murcell
1077 Janet Heath
1078 Mrs B. Ellicott
1079 Miss M. Bradby
1080 J.E. Park
1081 Mrs E.M. Mitchell-Fox
1082 Miss P.S. O'Flynn
1083 Miss M.G. Walder
1084 Mrs P.M. Hewson
1085 Dr Eric Morton
1086 Mrs Luff
1087 Mrs F.M. Hills
1088 Mr & Mrs M.I. Dove
1089 J.W. Miners
1090 James H. Hudson
1091 Miss B.A. Rix
1092 Miss J.P. Baker
1093 Mrs E. Pascoe
1094 Mr & Mrs M.G. Berryman
1095 Mr & Mrs C. Newby
1096 Mrs S.M. Hutchings
1097 R.A. Ledgerton
1098 Mrs V.S. Palmer
1099 J.M. Hughes
1100 D.W. Last
1101 Mr & Mrs M.G. Dunford
1102 Miss B.M. Dale
1103 Arthur R. Bray
1104 Peter Philp
1105 Mrs M.J. Morriss
1106 Rita Barkhuysen
1107 Marian J. Kallaway
1108 Mrs B.K. Clark
1109 Derek Jenkins
1110 Margaret Willis
1111 Brigid Daniel
1112 Gloria May
1113 Mrs C.L. Sly & Miss E. Sly
1114 Mr & Mrs G.H. Wood
1115
1116 Mrs R. Roskilly
1117 Miss M. Presland
1118 Mr & Mrs B. Wood
1119 Kevin P. Smith
1120 Mrs B.A. Coad
1121 Mrs S.M. Sage
1122 Mrs W.B. Stephens
1123 R.T. Brock
1124 Alexander James Shearman
1125 Miss Elizabeth A. Shearman
1126 Mrs V.H. Vittle
1127 J.B. Stanier
1128 Mrs J.V. Thompson
1129 Mrs Beryl White
1130 Miss S.G. Pomeroy
1131 Miss F.A. Hosier
1132 Mrs D.J.P. Jackson
1133 John L. Rapson
1134 David G. Burleigh

1135 Mrs R. Sherwood
1136 Mrs C. Diggle
1137 Mrs S. Lightbody
1138 Mrs F. Barriball
1139 W.G. Parker
1140 Paul A. Dowden
1141 Mr & Mrs J.F. Walters
1142 Miss M.C. Chambers
1143 B.J. Pengelly
1144 J.N. Edwards
1145 Mrs A.S. Cox
1146 Jeanne Bridge
1147 Maj T.F. Ellis OBE
1148 Mrs N. Abbey
1149 A.N. Haddy
1150 Mrs I.M. Baker
1151 G. Masters
1152 Mrs J.L. Ingram
1153 Mrs C.M. Taylor
1154 Miss N.M. Godson
1155 R.L. Rogers
1156 Steven Gook
1157 Mrs A.K. Stedham
1158 B. Mitchell
1159 B.R. Allen
1160 J.E. Packer
1161 P.J. White
1162 Mrs D.M. Treloar
1163 Michael Dorey
1164 Rev S.E.A. Underhill
1165 H. Broadbent
1166 Mrs Warlow-Harry
1167 Mrs J. Palmer
1168 Alan Rowland
1169 Mrs F.E. Lott
1170 Mrs R. Saltern
1171 Mrs G.V. Clark
1172 Mrs V.J. Burow
1173 Mrs A. Cleworth
1174 Mr & Mrs C.A. Jewell
1175 A.C. Cattermole
1176 Mr & Mrs Brian Adams
1177 Donald A. Blizzard
1178 Mrs Stratton
1179 J.A. Spiers
1180 Ian F. Long
1181 Dr Rosemary Lane
1182 Ursula Meldon
1183 Mrs D. Giles
1184 L.J.C. Bellingham
1185 Miss Jacka
1186 Dr Davy
1187 J.T. Williams
1188 Ronald Pye
1189 R.K. Barlow
1190 P.J. King
1191 Geoffrey C. Richards
1192 P. Billing
1193 J.R. Howes
1194 Walter & Hilary Grose
1195 Mr & Mrs B. Wareham
1196
1197 Mr & Mrs H.G. Woodhouse

1198 G.C. Matthews
1199 Mrs Heather Clark
1200 N. Stroud
1201 J. Peter Bray
1202 Mrs J.E. Hill
1203 Mrs Joan Beckingham
1204 Mrs J. Scott
1205 C.C. Marten
1206 Mr and Mrs G.A. Gibb
1207 Mrs E.F. Dalton White
1208 Mr & Mrs D.J. Hardman
1209 D. Hocking
1210 Mr & Mrs C. Uren
1211 Miss B. Strongman
1212 Elaine Hawking
1213 Andrew Perry
1214 Sandra Russell
1215 James Potter
1216 Mike Freemantle
1217 P.C. Firth
1218 R.I. Herbert
1219 Dr M. Kent

1220 Mrs C.W. Sharples
1221 Beryl Leverton-Simms
1222 Mrs Y. Wrigley
1223 Mrs Vera M. Morgan
1224 W. Cdr & Mrs H.R. Kerr
1225 Mrs Pauline Field
1226 E.M. Morgan
1227 G.C. Bowen
1228 Denone Rashleigh Richardson
1229 Mark Bawden
1230 I.J. Herring
1231 Miss P.C. Stone
1232 Mr & Mrs D.J. Gillingham
1233 Mr and Mrs J.J. Stewart
1234 D.A. Trevanion
1235 Mr & Mrs R.M. Nott
1236 Kenneth & Margaret Bawden

1237 C.M. Gardner
1238 Alison Roberts
1239 Lt Col W.G. Petherick
1240 Ivan Bickford
1241 Mrs B.M. Grayley
1242 Peter Parrott
1243 Mr & Mrs D.W. Holland
1244 Mrs P.J. Tempest
1245 Mr & Mrs J.W.M. Graham
1246 Valerie D. Brockenshire
1247 Ian D. Brokenshire
1248 D.S. Grose
1249 Miss Kathleen Eidmans
1250 Dr & Mrs M.J. Cotton
1251 V.L. Bowles
1252 Andrew Cleave
1253 Michael & Alix Lord
1254 D. Kirk
1255 James Edward Cousins

1256 K.M. Bowers
1257 A.G. Briggs
1258 Avice R. Wilson
1269 Mr & Mrs H.J. Boyd
1260 William Best Harris
1261 Valerie Vogtmann
1262 T.G. Storey
1263 O.C.A. Cornelius
1264 David Shone
1265 Kathleen Alexander
1266 Noel Jackson
1267 Cornelius Young
1268 J.A. Foskett
1269 Judyth Platt
1270 John C.L. Phillips
1271 John & Judith Gerrard
1272 Mrs Anne G. York
1273
1307 Cornwall County Library
1308 Miss V.W. Sulivan
1309 Mounts Bay School

Remaining names unlisted